T0349043

Radar Imaging of the Ocean Waves

Radar Imaging of the Ocean Waves

Mikhail B. Kanevsky

Institute of Applied Physics
Russian Academy of Sciences

ELSEVIER

Amsterdam • Boston • Heidelberg • London • New York • Oxford
Paris • San Diego • San Francisco • Singapore • Sydney • Tokyo

Elsevier
Linacre House, Jordan Hill, Oxford OX2 8DP, UK
Radarweg 29, PO Box 211, 1000 AE Amsterdam, The Netherlands

First edition 2009

British Library Cataloguing in Publication Data
A catalogue record for this book is available from the British Library

Library of Congress Cataloging-in-Publication Data
Kanevsky, M. B. (Mikhail Borisovich)
  Radar imaging of the ocean waves / Mikhail. B. Kanevsky. — 1st ed.
    p.   cm.
  Includes bibliographical references and index.
  ISBN 978-0-444-53209-1
  1. Ocean waves—Remote sensing.   2. Doppler radar.   3. Synthetic aperture radar.   I. Title.
  GC211.2.K359   2009
  551.46'3—dc22

                                                                    2008031146

For information on all Elsevier publications
visit our website at www.elsevierdirect.com

Printed and bound in the United Kingdom
Transferred to Digital Printing, 2010

Working together to grow
libraries in developing countries

www.elsevier.com | www.bookaid.org | www.sabre.org

ELSEVIER   BOOK AID
            International   Sabre Foundation

Contents

Introduction

When discussing on the radar as a tool for remote sensing of the ocean, the main emphasis is on its indisputable benefits such as all-weather capability and wide swath. Indeed, microwave radar mounted on a space platform sees the ocean surface through the atmospheric layers at any time of the day and year in the swath over 100 km wide; thus, in a relatively short time span, the radar manages to survey all the oceans in the world.

However this book does not aim at covering the range of radar capabilities again – this was done in great detail by many other works. As a fine example we can cite the excellent book *Synthetic Aperture Radar Marine User's Manual*, published by the US Department of Commerce and National Oceanic and Atmospheric Administration in 2004, and supplied with a number of remarkable illustrations. The author's goal is to give the basics of the ocean surface radar imaging theory, exemplify it with experimental data and point out the pitfalls created by the specific nature of radar imaging mechanisms. Neglect of the specifics can lead (and has occasionally led) to seriously erroneous interpretation of ocean surface radar images.

Radar oceanography can be traced back to early 1950s of the twentieth century, when experiments (Crombie 1955, Braude 1962) were first conducted with high-frequency (HF) coastal radars where the sea was regarded as a study object rather than the source of perturbations or interferences by locating various surface and near-surface targets. The experiments have initiated a multinational comprehensive theoretical and experimental study of the HF, VHF (very high frequency) and microwave scattering by the ocean surface with coastal, ship-borne and airspace radars.

In 1978 radio oceanography gained momentum due to the launch of an oceanographic satellite SEASAT (USA) with an extensive inventory of radar equipment on board, including synthetic aperture radar (SAR). It was followed by a whole series of over a dozen of SAR-based satellites put into space by different countries and international organizations.

Numerous publications featuring ocean surface images, which cover various ocean and atmospheric processes disclose little against the ample data provided by space SARs. However, a fair qualitative interpretation of SAR ocean images requires an accurate knowledge of the mechanism (or mechanisms) which allow SAR to see the waves. SAR ocean images are often perceived and interpreted as if they have been made with conventional incoherent side-looking radar characterized by hypothetical super-high resolution. That is absolutely erroneous! The point is that standard aperture synthesis procedure incorporates reflected signal phase history, corresponding to the static surface while the water surface continuously moves randomly. What is the connection then between roughness spectrum and the one of the respective SAR images? What is, on the whole, SAR performance potential in ocean surface global monitoring, particularly in the coastal regions where the need for a higher resolution against other space radars is due to significant spatial roughness variability? A substantial part of the book is devoted to these questions.

These are the rough water surface imaging mechanisms that the present research focuses on. We do not dwell on the physical aspects of the surface, sub-surface and atmosphere processes which somehow modify roughness characteristics and through these modifications are manifested on the surface. Various manifestations of dynamic "ocean-to-atmosphere" system processes in ship-borne and airspace radar ocean images are covered in multiple works which are not going to be surveyed herein. Nevertheless, understanding the influence of a certain physical process on the roughness characteristics supported by the theory presented herein helps elicit its manifestation character and degree in radar imagery of the ocean surface.

Chapters 1, 2 and Chapter 3, partly, are introductory. They provide an insight into the basis of the material to follow. Despite the number of works on rough surface scattering theory and its experimental evidence we deemed it expedient to brief the reader on the relevant issues, which helps him to get down to the theory of radar imaging without reference to supplementary sources. Besides, Chapter 3 also contains some original information, for example, on the method of spotting slicks on the sea surface by the backscattered signal Doppler shift when probing at intermediate (between small and moderate) incidence angles.

Chapter 4 presents the theory of Doppler spectrum of microwave signal backscattered from the sea surface at moderate incidence angles. The most solid criterion for the theory, Doppler spectrum is also the key scattering characteristics containing almost exhaustive information on the scattering surface statistic parameters.

Chapter 5 expounds on the theory of sea imaging via incoherent side-looking real aperture radar (RAR). Strictly speaking, in RAR case both roughness imaging mechanisms and accompanying the speckle noise are known. However they used to be treated separately; therefore, it seems reasonable to combine them into a well-structured theory, where they are implied by a single primary formula for a returned signal realization. This approach helps compare energetic value of various signal fluctuations responsible for imaging and speckle-noise formation. The chapter also examines the way RAR sees roughness agitated by large-scale intra-ocean processes, particularly internal waves.

Chapter 6 dwells on SAR. It develops a consistent theory of ocean SAR imaging. Based on the composite model of microwave scattering by rough water surface presented in Chapter 3, the theory describes both roughness imaging and speckle noise, which hampers significantly ocean image interpretation. A new spectral estimate is offered for a statistically speckle-noise-free image spectrum retaining SAR resolution.

Ocean roughness imaging mechanism receives a special attention. Analysis and numeric modelling have shown that the scatterers' effective density fluctuation mechanism works in fact solely if there are rather flat swell and no large-scale wind waves. With wind waves another mechanism governs the process, namely, surface element number fluctuations, which images shift randomly due to orbital velocities and superimpose in the image plane. SAR signal intensity fluctuations are mainly associated exactly with these above fluctuations. It has been shown that the well-known spectral cut-off is nothing but the manifestation of above-mentioned imaging mechanism in the spectral domain.

Finally, in Chapter 7 we examine some aspects of advanced (interferometric and polarimetric) radars used for remote sensing of the ocean surface.

This book, which is a fruit of the author's prolonged work in the area of radar oceanography, would not be possible without a close collaboration with Dr V.Yu. Karaev, Mr L.V. Novikov and Mrs. G.N. Balandina. At different stages of the work the data covered by the book have been discussed with participation of Prof. W.R. Alpers, Prof. L.S. Dolin and Dr W.L. Weber. The author expresses his deepest gratitude to all of them.

The author is grateful to Prof. L.A. Ostrovsky, as well as to Editors of Elsevier Monograph Series Prof. G. Zaslavsky and Dr A. Lou for their help in the book publication.

The author especially appreciates the assistance rendered by Alyona Kanevskaya when writing the manuscript in English.

– 1 –

Preliminary notes on radar imaging

In this short introductory chapter, some notions on radar imaging are given to introduce the reader to the means and methods of radar imaging of the ocean surface. Those who are familiar with the basics of radar operation can skip the chapter; it will not hamper the comprehension of further text.

Figure 1.1 presents the surveying scheme for the earth (ocean) surface drawn for side-looking microwave radar installed on the platform moving in the plane ZY parallel to the axis Y.

The radar operates in the pulse mode, and each emitted pulse every instant makes an illuminated patch on the surface (shown in Figure 1.1 as a cross-hatched section), moving with the speed of light from the near edge to the far one of the swath; the swath size L is determined by the angular width of radar antenna pattern in vertical plane. Due to the small ratio of the azimuthal (along the Y-axis) size of the patch to its range, the ground range boundaries are marked as rectilinear. In fact, these boundaries appear as two arcs of concentric circles centred at the point of the radar antenna centre projection on the Y-axis.

As shown in Figure 1.1, the azimuthal size of this patch $\Delta y = R\delta_y$, where R is the slant range of the patch central point and δ_y is the antenna pattern width in the azimuthal plane.

This section is called radar resolution cell, as the points on the surface within its bounding are indiscernible. The indiscernibility and the ambiguity associated with it are obvious for the pair of rays with an angular distance less than δ_y, which have been backscattered by the same spot. As for the range resolution, the ambiguity occurs for the signals reflected by surface spots at different distances but received by the antenna simultaneously. This is the case when one of the signals has been transmitted later than the other, but has returned from the nearer point. The ground range size Δx of the resolution cell is found on the condition that leading and trailing pulse edges, scattered by different points on the surface, reach back the antenna synchronously. In Figure 1.2 there are two rays in the vertical plane bordering the resolution cell, the radar antenna supposedly being in the far zone. The leading edge of the pulse with the duration τ_p reflected by the surface

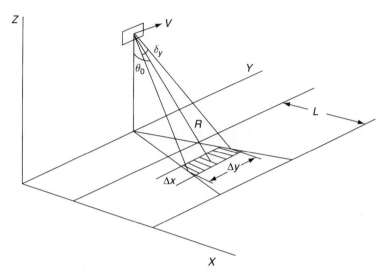

Figure 1.1 Radar probing geometry.

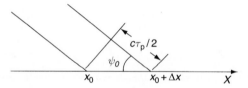

Figure 1.2 Formation of radar resolution cell range scale.

point $x_0 + \Delta x$ has to cover the additional distance $c\tau_p/2$ twice to reach the antenna simultaneously with the trailing edge backscattered from the point x_0 (here c is the speed of light). Geometrically,

$$\Delta x = \frac{c\tau_p}{2\cos\psi_0} \tag{1.1}$$

where $\psi_0 = \pi/2 - \theta_0$ is the grazing angle. Thus, the shorter the transmitted pulse, the higher the ground range resolution.

However when working at large distances, where a powerful emitted signal is essential, it is technically more expedient to deploy long pulses with linear frequency modulation (chirp pulse). It requires Doppler radar which affords to keep track of the signal phase. Inside the pulse the emission frequency changes according to $\omega = \omega_0 + \gamma t$ within the limits of $\omega_0 \pm \gamma\tau_p/2$; the value $\gamma\tau_p/2$ is termed frequency deviation. Each chirp pulse after its reception is subjected to compression; that is to say its complex envelope undergoes matched filtration. As it is known, the matched filtration operation is mathematically described as the convolution

$$F(t) \propto \int dt' f(t') f^*(t - t') \tag{1.2}$$

where the asterisk denotes complex conjugation. After the filter is adjusted to the chosen range R (this range corresponds to the return time $t = 2R/c$ of the emitted electromagnetic field) we obtain for the pulse complex envelope (or, which is the same, complex amplitude),

$$a(t) \propto \int_{-\tau_p/2}^{\tau_p/2} dt' \exp\left(i\gamma t'^2\right) \exp\left\{-i\gamma\left[\left(t - \frac{2R}{c}\right) - t'\right]^2\right\} \propto \frac{\sin\left[\gamma(t - 2R/c)\tau_p\right]}{\gamma(t - 2R/c)\tau_p}$$

$$\tag{1.3}$$

As we see, the pulse duration τ'_p after matched filtration turns out to be proportional to the frequency deviation and can be considerably compressed. As a result we get a compressed pulse, and the range resolution is now determined by the compressed pulse duration $\Delta x = c\tau'_p/2\cos\psi_0$. The described procedure physically means that over the whole period while the long chirp pulse passes through the point with the given range, matched filtration provides extraction and accumulation of the backscattered signal part reflected exactly by this point (more exactly, its vicinity area whose size is determined by the frequency deviation).

The azimuth resolution conventional real aperture radar (RAR) is restricted by the diffraction limit $\delta_y = \lambda/D$, where δ_y is the one-way path antenna pattern angular width determined at half-power level, λ is the radar wavelength and D is the antenna extension along the Y-axis. Obviously, space radar survey (in the range about thousand of kilometers) would require building and mounting in space a huge antenna in order to achieve a sufficiently high azimuthal resolution. Thus, for instance, for $\lambda = 3$ cm and $R = 1000$ km, to obtain the resolution $(\lambda/D)R = 10$ m we might have the antenna with azimuthal extension $D = 3$ km. Hence at large distances the so-called synthetic aperture radars (SARs) are deployed, which at a rather small antenna size provide high azimuthal resolution by means of coherent procession of Doppler signal.

Let us examine Figure 1.1, assuming that the ocean surface is surveyed with SAR. The high resolution over the azimuthal coordinate y is achieved by means of coherent procession with the help of

$$a_{\text{SAR}}(t) = \frac{1}{\Delta t} \int_{t-\Delta t/2}^{t+\Delta t/2} dt' \, a(t') \exp\left[-i\frac{k}{R}V^2(t'-t)^2\right] \tag{1.4}$$

Here a and a_{SAR} are the complex amplitudes of reflected and synthetic signals, and Δt is the SAR integration time; the multiplier before the integral serves to preserve dimensionality.

To understand Eqn. (1.4) in depth, we are going to temporarily switch from the yet insignificant dependence of physical values on x to an image line along the Y-axis. For the slant range R' of arbitrary point y' one can write:

$$R' \approx R + \frac{(y'-Vt')^2}{2R} \tag{1.5}$$

In this approximation, the Doppler phase of the signal coming from y' looks like

$$\Phi_{\mathrm{D}} = 2kR + \frac{k}{R}(y'-Vt')^2 \tag{1.6}$$

For the distribution of the incidence field amplitude on the surface, we adopt a simple approximation

$$\phi\left(\vec{r}'-\vec{r}\right) = \begin{cases} 1, & \vec{r}'-\vec{r} \in \Delta\vec{r} \\ 0, & \vec{r}'-\vec{r} \notin \Delta\vec{r} \end{cases} \tag{1.7}$$

where the rectangular shaped area $\Delta\vec{r}\,(\Delta x, \Delta y_1)$ is the physical resolution sell segment used for aperture synthesis, in which the antenna pattern main lobe provides nearly homogeneous distribution of the incidence field amplitude. Then we can write for the complex amplitude of the backscattered field,

$$a(t') \propto \mathrm{e}^{2\mathrm{i}kR} \int\limits_{\Delta y_1} \mathrm{d}y' p(y') \exp\left[\mathrm{i}\frac{k}{R}(y'-Vt')^2\right] \tag{1.8}$$

where $p(y')$ is the reflection coefficient of the surface, which so far has been assumed to be static.

We choose a point $y' = Vt$ on the surface that we would like to image. According to Eqn. (1.8), the backscattered signal changes resulting from the phase modulation caused by radar motion are described by the multiplier

$$\mathrm{e}^{\mathrm{i}\Phi(t')} = \exp\left[\mathrm{i}\frac{k}{R}V^2(t-t')^2\right] \tag{1.9}$$

Now let us return to expression (1.4). Clearly, multiplying $a(t')$ by $\mathrm{e}^{-\mathrm{i}\Phi(t')}$ and integrating over t', we first offset the phase modulation of the signal returned from the point $y' = Vt$, and second over the period Δt accumulate the signal *in phase* unlike the signals backscattered by other points. Therefore, transformation (1.4) is nothing but a matched filtration operation to recover the part that has been returned from the point $y' = Vt$ from the backscattered field.

Notably, according to Eqn. (1.6),

$$\omega_{\mathrm{D}} = \frac{d\Phi_{\mathrm{D}}}{dt'} = -2\frac{kV}{R}(y'-Vt) \tag{1.10}$$

which is the linear change of the Doppler frequency of the backscattered signal. The latter thus turns out to be a chirp signal, i.e., a signal with the linear frequency modulation caused by radar movement. Transformation (1.4) here proves analogous to the signal compression.

In other words, SAR processing entails the correlation of the return signal phase with that of a suitable reference signal (see Eqn. (1.9)). If the reference signal phase involves many cycles, this is called a focused synthetic aperture. The formation of a synthetic aperture can be simplified by using a shorter history of return signal and a corresponding reduction in reference signal phase change; in this case the azimuth resolution will degrade. In the limit, the simplest synthetic aperture can be formed with a constant (more exactly, almost constant) reference signal phase and a relatively short history of returned signal

$$\frac{k}{R}V^2(t-t')^2 < \frac{\pi}{2} \tag{1.11}$$

and, consequently,

$$V\Delta t < \sqrt{\lambda R} \tag{1.12}$$

This is called an unfocused synthetic aperture.

The azimuthal size of SAR resolution cell in view of the linear relation between a_{SAR} and $p(y')$ can be found through impulse characteristics (or point-target response) of SAR as a linear filter; we obtain it by assuming $p(y') \propto \delta(y' - y_0)$ in Eqn. (1.8):

$$I_{\mathrm{SAR}}(t) = |a_{\mathrm{SAR}}(t)|^2 \propto \left(\frac{\sin u}{u} \right)^2 \tag{1.13}$$

$$u = \frac{kV\Delta t}{R}(y_0 - Vt)$$

It shows that in the radar image plane, the azimuthal size of the spot resulting from the point scatterer (i.e., over two-way path of radar pulse) at the level of twofold power decrease is

$$\Delta_{0,\mathrm{SAR}} = \frac{\lambda R}{2V\Delta t} \tag{1.14}$$

This is the well-known SAR nominal resolution, which, as we shall further see, differs from the real one due to the peculiar nature of SAR survey over agitated water surface. Considering that it is pointless to set SAR integration time value Δt higher than the presence period of the scatterer in the illuminated spot, i.e. $\Delta t \leq \lambda R / DV$ (where D is the antenna azimuthal size), it turns out that $\Delta_{0,\mathrm{SAR}} \geq D/2$ whatever the distance is, at the same time for RAR:

$$\Delta_{\mathrm{RAR}} = \frac{\lambda}{D} R \tag{1.15}$$

Comparing Eqns (1.14) and (1.15), we see that Eqn. (1.4) equals to building an antenna with the linear azimuthal size $D_{\mathrm{SAR}} = V\Delta t$, and multiplier 2 in denominator (1.14) implies that here unlike Eqn. (1.15) the azimuthal size is calculated over two-way path.

Yet, as an ocean surface sensing tool, SAR is not by far analogous to incoherent side-looking radar with a hypothetically superhigh resolution. Rather a complicated issue of the ocean roughness SAR image structure has long been a controversial subject (see, for instance, the reviews Hasselmann et al. (1985) and Kasilingam and Shemdin (1990)). The issue is the real phase history of the signal returned from the surface stationary element differs considerably for the same element randomly moving under the rough-sea conditions compared to pattern (1.9) of a reference signal phase history – the groundwork of the matched filtration operation (1.4). It is explained in greater detail in Chapter 5.

Surface imaging imposes certain conditions on pulse repetition frequency (PRF). Evidently, the pulses should be spaced apart in time to avoid their overlapping or mixing up, i.e. the situation when the preceding pulse backscattered from a greater distance reaches the antenna simultaneously or even after the subsequent pulse reflected at a smaller distance. This requirement can be expressed by

$$T_{\mathrm{p}} > \frac{2L}{c} \quad \text{or} \quad \mathrm{PRF} < \frac{c}{2L} \tag{1.16}$$

where T_{p} is the pulse repetition period, and L is the swath width. On the contrary, PRF has to be high enough not to lose the information contained in the backscattered signal. The minimum frequency meeting the requirement is called Nyquist frequency and equals twice the value of the maximum frequency in the reflected field Doppler spectrum.

The physics of the Nyquist frequency is quite transparent and is as follows. The information in the reflected signal is in the spectrum of its complex envelope, i.e. in the signal Doppler spectrum. The envelope clearly cannot undergo any significant changes during the time span shorter than half the period corresponding to the maximum frequency $f_{\mathrm{D}}^{\max}$ in its spectrum. Consequently, sampling the Doppler signal over intervals $1/2f_{\mathrm{D}}^{\max}$, we will not miss the useful data. Therefore PRF will satisfy the relation

$$\mathrm{PRF} \geq 2f_{\mathrm{D}}^{\max} = F_{\mathrm{Nyquist}} \tag{1.17}$$

As a result, the PRF deployed in imaging radars proves to be limited as to its minimum and maximum values, i.e.

$$\frac{c}{2L} > \mathrm{PRF} > 2f_{\mathrm{D}}^{\max} \tag{1.18}$$

We further find f_D^{max} in the following way. The Doppler frequency of the returned signal is given by

$$f_D = \frac{2V_{rad}}{\lambda} \qquad (1.19)$$

where $V_{rad} \approx V\varphi$ is the radial velocity of a surface element surveyed at the azimuthal angle φ from the antenna dislocation place. It is obvious that the maximum Doppler frequency is characteristic of the signal received by the leading along-track edge of antenna pattern in the azimuthal plane near to its null. This edge ray in view for the two-way path is directed at the angle

$$\varphi = \frac{\lambda}{2D} \qquad (1.20)$$

against the broadside axis of the antenna pattern; hence

$$f_D^{max} = \frac{V}{D} \qquad (1.21)$$

Consequently,

$$\frac{c}{2L} > PRF > \frac{2V}{D} \qquad (1.22)$$

In case of space SAR the backscattered signal is characterized by broad Doppler spectrum as a result of high values of the radar platform velocity; therefore, condition (1.22) is rather stringent, yet satisfiable.

Note that Eqn. (1.22) should be used only in preliminary calculations; in actual design the Earth curvature should be taken into account.

Space SAR operation, design, technical characteristics and deployment specifics are described in greater detail in the tutorial articles (Tomiyasu 1978, Elachi et al. 1982).

The energy characteristics of backscattered radar signal there usually serves the concept of radar cross section σ or normalized radar cross section (NRCS) $\sigma_0 = \sigma/S_0$, where S_0 is the scattering patch area. We will further introduce the concept.

Let the radar at the range of R from the scattering patch emit the power equal to P_t. With isotropic emitter the power will be spread over the spherical surface, and close to the mentioned patch its density will be $P_t/4\pi R^2$. Because a realistic radar antenna pattern is anisotropic, the power density should be multiplied by the antenna gain G. Then the radiated power is being collected by the patch having the effective square σ, and after that it is being backscattered by this patch. Finally, the backscattered field with power density is reduced again by $1/4\pi R^2$ and is being collected by a radar antenna with an effective area A_e. Thus, the received power is

$$P_r = \left(\frac{P_t}{4\pi R^2}G\right)\sigma\frac{1}{4\pi R^2}A_e \qquad (1.23)$$

Theoretical works as a rule skip radar antenna specifications (they can be described a posteriori), and assuming that the radar antenna is located far from the scattering patch, suppose

$$\sigma_0 = \frac{4\pi R^2}{S_0} \lim_{R \to \infty} \frac{P_r}{P_t/4\pi R^2} \tag{1.24}$$

or

$$\sigma_0 = \frac{4\pi R^2}{S_0} \lim_{R \to \infty} \frac{|E_r|^2}{|E_0|^2} \tag{1.25}$$

where E_0 is the incident field near the scattering surface and, therefore,

$$\frac{|E_r|^2}{|E_0|^2} = \frac{P_r}{P_t/4\pi R^2} \tag{1.26}$$

For this reason, when speaking of NRCS of the ocean surface, we shall further refer to the value defined by Eqn. (1.25).

– 2 –

Description of the sea surface

This chapter briefly covers the basics of roughness that matter for the radar imaging of the ocean surface.

Ocean roughness is caused by air–sea interaction and at the same time it plays an important role in this interaction. When the wind blows over the ocean, a turbulent airflow first incites short waves of centimetre wavelengths (centimetre ripples) through friction stresses. These waves transmit part of the wind energy to the longer waves (by means of weak non-linear wave–wave interaction); the other part is dissipated due to viscosity. Later on, the longer waves interacting with the wind take over its energy and communicate it on to the more longer waves through weak non-linear interaction. Thus with further time the wave energy focused originally in the ripple range grows into increasingly large-scale waves tens and hundreds meters long, which are commonly associated with sea roughness. Large wave growth goes on until the balance is reached when the wind-generated energy becomes equal to the energy dissipated by wave breaking, turbulence and viscosity.

As a result of this the ocean surface develops roughness of various sizes, and one can see at a glance that roughness description work is unfeasible without statistical analysis methods.

2.1 SEA WAVE SPECTRA: GENERAL RELATIONSHIPS

The facts described above display that sea roughness is generally speaking non-stationary and inhomogeneous in space. The spatio-temporal scale though of this homogeneity and stationarity in the open ocean (i.e. outside the coastal area) is fairly large – approximately 50 km and 1 h, respectively. That is why the roughness field can be considered as statistically homogeneous and stationary within the given above spatio-temporal interval.

Consider the Cartesian coordinates system with a horizontal plane XY corresponding to the average level of the rough ocean surface and analyse the statistical properties of the function $z(x,y,t)$, describing surface elevations above the average level $z = 0$.

If we see the surface $z(x,y,t)$ as a superposition of numerous running sinusoidal waves with different lengths, amplitudes and phases (Longuet-Higgins 1952), then by the central limit theorem of the probability theory the surface elevation probability density function

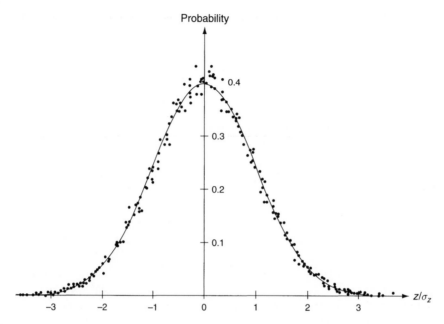

Figure 2.1 Probability density function of sea surface elevation (solid line) and the experimental points (Stewart 1985; originally cited from Carlson et al. 1967).

(PDF) will be rather close to the Gaussian one, which is in good conformity with experimental data. Thus the PDF of sea surface elevation, normalized by the standard deviation $\sigma_z = \langle z^2 \rangle^{1/2}$ (see Figure 2.1 quoted from Stewart (1985), originally given in Carlson et al. (1967)), testifies that the observations are well fit by the Gaussian curve.[1] Therefore the correlation function

$$B_z(\vec{\rho}, \tau) = \langle z(\vec{r}, t) z(\vec{r} + \vec{\rho}, t + \tau) \rangle \tag{2.1}$$

yields practically a full statistical description of the roughness field. Here $\vec{r} = \{x, y\}$ is the radius vector of an arbitrary point on the plane $z = 0$, $\vec{\rho}$ and τ are spatial and temporal shifts, respectively, and angle brackets stand for statistical averaging. The dependence of B_z solely on $\vec{\rho}$ and τ (on $|\tau|$, to be more exact; however hereafter we will omit the modulus sign where appropriate for simplification) is caused by the assumed statistical homogeneity and stationarity of the field $z(\vec{r}, t)$.

---

[1] To be more exact, the hypothesis of the roughness Gaussian statistics should be accompanied by certain allowances, which is accounted in particular by the rather vivid tendency of ocean waves to have asymmetrical profiles as well as rougher crests and smoother troughs, than is defined by the Gaussian statistics. However, we will build our further calculations on the Gaussian PDF, assuming that this adjustment does not considerably affect the results.

As is known, the correlation function B_z is related to the spatio-temporal spectrum Ψ_z through the Wiener–Khintchin correlation:

$$B_z(\vec{\rho}, \tau) = \iint d\vec{\kappa}\, d\omega\, \Psi_z(\vec{\kappa}, \omega) \exp\left[i(\vec{\kappa}\vec{\rho} - \omega\tau)\right] \tag{2.2}$$

$$\Psi_z(\vec{\kappa}, \omega) = \frac{1}{(2\pi)^3} \iint d\vec{\rho}\, d\tau\, B_z(\vec{\rho}, \tau) \exp\left[-i(\vec{\kappa}\vec{\rho} - \omega\tau)\right] \tag{2.3}$$

where $\vec{\kappa} = \{\kappa_x, \kappa_y\}$ is the spatial wave number and ω is the temporal frequency. Spatial $\hat{W}_z(\vec{\kappa})$ and temporal $\hat{G}_z(\omega)$ spectra naturally follow from $\Psi_z(\vec{\kappa}, \omega)$:

$$\hat{W}_z(\vec{\kappa}) = \int d\omega\, \Psi_z(\vec{\kappa}, \omega) \tag{2.4}$$

$$\hat{G}_z(\omega) = \int d\vec{\kappa}\, \Psi_z(\vec{\kappa}, \omega) \tag{2.5}$$

which gives for elevation variance $\sigma_z^2 = \langle z^2 \rangle$

$$\sigma_z^2 = \int d\vec{\kappa}\, \hat{W}_z(\vec{\kappa}) = \int d\omega\, \hat{G}_z(\omega) \tag{2.6}$$

We represent $z(\vec{r}, t)$ as the complex Fourier integral:

$$z(x, y, t) = \iint d\vec{\kappa}\, d\omega\, A_z(\vec{\kappa}; \omega) \exp[i(\vec{\kappa}\vec{r} - \omega t)] \tag{2.7}$$

Since value z is the real one, the following holds true for the spectral amplitude A_z:

$$A_z(\vec{\kappa}, \omega) = A_z^*(-\vec{\kappa}, -\omega) \tag{2.8}$$

where the asterisk denotes the complex conjugation, and due to the presumable statistical homogeneity and stationarity of the roughness field,

$$\langle A_z(\vec{\kappa}, \omega) A_z^*(\vec{\kappa'}, \omega') \rangle = \Psi_z(\vec{\kappa}, \omega) \delta(\vec{\kappa} - \vec{\kappa'}) \delta(\omega - \omega') \tag{2.9a}$$

$$\langle A_z(\vec{\kappa}, \omega) A_z(\vec{\kappa'}, \omega') \rangle = \Psi_z(\vec{\kappa}, \omega) \delta(\vec{\kappa} + \vec{\kappa'}) \delta(\omega + \omega') \tag{2.9b}$$

Taking into account Eqn. (2.9) and also Eqns (2.4) and (2.5), we can obtain from Eqn. (2.7) the expressions for the surface slope variance:

$$\sigma_{sl}^2 = \left\langle \left(\frac{\partial z}{\partial x}\right)^2 + \left(\frac{\partial z}{\partial y}\right)^2 \right\rangle = \int d\vec{\kappa}\, \kappa^2 \hat{W}_z(\vec{\kappa}) \tag{2.10}$$

and for the variance of the surface vertical velocity

$$\sigma_{\text{vert}}^2 = \left\langle \left(\frac{\partial z}{\partial t}\right)^2 \right\rangle = \int d\omega\, \omega^2 \hat{G}_z(\omega) \tag{2.11}$$

The spectra $\hat{W}_z$ and $\hat{G}_z$ are characterized respectively by central symmetry (symmetry relatively the origin) and parity (evenness), i.e.

$$\hat{W}_z(\vec{\kappa}) = \hat{W}_z(-\vec{\kappa}), \quad \hat{G}_z(\omega) = \hat{G}_z(-\omega) \tag{2.12}$$

Taking into account the parity of the spectrum $\hat{G}_z$, we have

$$\sigma_z^2 = 2\int_0^\infty d\omega\, \hat{G}_z(\omega) = \int_0^\infty d\omega\, G_z(\omega) \tag{2.13}$$

$$\sigma_{\text{vert}}^2 = \int_0^\infty d\omega\, \omega^2 G_z(\omega) \tag{2.14}$$

As for the spatial spectrum $\hat{W}_z$, at first sight, the central symmetry seems to contradict to the obvious intensity difference between the downwind and upwind waves, although in effect there is no discrepancy here. The matter is that the vector $\vec{\kappa}$ itself does not define the wave's propagation direction, which also depends on the temporal frequency sign. Thus, for the wave $w \propto \exp\left[i(\vec{\kappa}\,\vec{r} - \omega t)\right]$ the vector $\vec{\kappa}$ and the wave propagation direction coincide at the positive frequency and are opposed when the frequency is negative. Therefore the pair $(-\vec{\kappa}, -\omega)$ corresponds to the same wave as $(\vec{\kappa}, \omega)$.

Restriction to only positive temporal frequencies allows us to introduce the spatial spectrum

$$W_z(\vec{\kappa}) = 2\hat{W}_z^+(\vec{\kappa}) \tag{2.15}$$

that lies within the boundaries of wave numbers corresponding to actual surface waves propagation vectors, so that

$$\sigma_z^2 = \int d\vec{\kappa}\, W_z(\vec{\kappa}) \tag{2.16}$$

$$\sigma_{\text{sl}}^2 = \int d\vec{\kappa}\, \kappa^2 W_z(\vec{\kappa}) \tag{2.17}$$

The study of spectra $\Psi_z(\vec{\kappa}, \omega)$ and $W_z(\vec{\kappa})$ is associated with certain technical obstacles, as it requires creating a large network of spaced-apart synchronized roughness-sensing devices; for these reasons, when it comes to practice, as a rule, the analysis of the frequency spectrum $G_z(\omega)$ is confined to a single point in space.

Numerous experiments have resulted in a fairly large number of empirical frequency spectrums of wind waves. All the spectra are characterized by certain common features, which are best seen in case of fully developed wind waves when under rather steady wind conditions input energy offsets dissipated energy at all wavelengths of the spectrum. According to Figure 2.2 (from Davidan et al. 1985), which displays a spectrum of a fully developed wind-induced roughness at the near-surface wind speed of $U = 10\,\text{ms}^{-1}$, these features are a sudden leap of spectral density from low frequencies to the peak frequency and its relatively gradual slide at the transition from the peak to more higher frequencies. Another inherent spectrum feature is the higher the wind speed the lower is the peak frequency.

Notably the weaker secondary maximum, hardly discerned in the frequency spectrum in Figure 2.2, corresponds to the lower boundary line of the so-called equilibrium sub-range. In terms of wave numbers, this sub-range is between about 1 and $10\,\text{rad}\,\text{m}^{-1}$; here the balance between the wind-induced energy (through the high wave number end) and the energy transmitted to the longer waves in the spectrum (through the low wave number end) takes place. That is why the spectral density does not depend much on the wind speed inside this sub-range. The secondary peak is not always distinctly pronounced; sometimes it is just a disturbance in an evenness of the spectral density (marked by a dotted line in Figure 2.2) (Davidan et al. 1985).

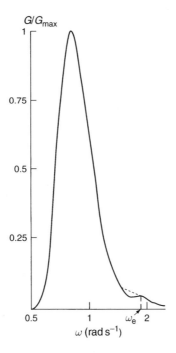

Figure 2.2 Frequency spectrum of fully developed windsea in the gravity area at the wind speed $U = 10\ \text{m s}^{-1}$ (Davidan et al. 1985).

Due to the restoring forces applied to the declined surface the whole spectrum is divided into three intervals: gravity (waves over 7 cm long), gravity–capillary (0.6–7 cm long) and capillary (under 0.6 cm). (It is well understood that the interval boundaries do not exactly lie within accurate precision figures, just as the application spheres of physical forces are not strictly marked off.) Respective borderline temporal frequencies are thus determined by the correlation given by the linear hydro-dynamic theory. This theory defines the surface as the sum of non-correlating waves which wave numbers and temporal frequencies are bound by the dispersion relationship:

$$\omega^2 = \kappa g \left(1 + \frac{\alpha_s}{\rho_w g} \kappa^2 \right) \tanh(\kappa D) \tag{2.18}$$

where α_s is the surface tension, ρ_w the water density, g the acceleration due to gravity, and D the water depth; $\alpha_s/\rho_w = 74 \, \text{cm}^3 \, \text{s}^{-2}$ for a clean water surface. For deep water $\kappa D \gg 1$; therefore $\tanh(kD) \approx 1$, and thus Eqn. (2.18) results in

$$\omega^2 = \kappa g \left(1 + \frac{\alpha_s}{\rho_w g} \kappa^2 \right) \tag{2.19}$$

In the gravity interval, which holds over 99% of the wave energy, the surface tension as a restoring force is insignificant against the gravity, and the dispersion correlation takes a very simple form:

$$\omega^2 = \kappa g \tag{2.20}$$

Based on Eqn. (2.19) the frequency borderlines of the spectral intervals are defined as

$$\frac{\omega}{2\pi} < 5\text{Hz (gravity interval)}$$

$$5\text{Hz} < \frac{\omega}{2\pi} < 48\text{Hz (gravity–capillary interval)}$$

$$\frac{\omega}{2\pi} > 48\text{Hz (capillary interval)}$$

It is worth noting that this differentiation is also largely conventional, first of all due to the fact that gravity–capillary and capillary waves are influenced by the gravity waves; thus the dispersion correlation (2.19) is generally speaking violated in small-scale area if large gravity waves are present.

2.2 GRAVITY WAVE SPECTRA

There is a well-known approximation for the gravity part of the fully developed roughness spectrum, namely, Pierson–Moskowitz spectrum (Pierson and Moskowitz 1964):

$$G_z(\omega) = 8.1 \times 10^{-3} g^2 \omega^{-5} \exp\left[-1.25 \left(\frac{\omega_m}{\omega} \right)^4 \right] \tag{2.21}$$

where ω_m is the maximum frequency, $\omega_m U/g = 0.83$, and the wind speed U is supposed to be measured at the height of $10\,m$. Note also that considering Eqn. (2.21), the integrals in Eqns (2.13) and (2.14) can be taken analytically; therefore, in case of fully developed windsea

$$\sigma_z = \left[\int_0^\infty d\omega\, G_z(\omega)\right]^{1/2} \approx 5.7 \times 10^{-2} U^2/g \qquad (2.22)$$

$$\sigma_{\text{vert}} = \left[\int_0^\infty d\omega\, \omega^2 G_z(\omega)\right]^{1/2} \approx 6.8 \times 10^{-2} U \qquad (2.23)$$

However, the fully developed windsea takes time during which the roughness spectrum undergoes certain changes before it attains the shape of Eqn. (2.21) some time after the start point of air–sea interaction. As the roughness progresses, the energy obtained from the wind is applied on to the more longer waves, the spectrum peak, more enhanced than in Eqn. (2.21), shifts to the more lower frequencies and the roughness energy, i.e. the integral of Eqn. (2.21), gathers momentum. This process includes quite an important parameter, wind fetch X_w – the distance over which the wind blows. In the idealized case of constant wind blowing away from a lee shore, this is the distance from the shore to the given area of the sea surface. We can find the position of the maximum in the developing windsea spectrum with an empirical correlation (Hasselmann et al. 1973):

$$\tilde{\omega}_m = 22\tilde{X}_w^{\;-0.33} \qquad (2.24)$$

where $\tilde{\omega}_m = \omega_m\, U/g$ and $\tilde{X}_w = X_w g/U^2$. The windsea is supposed to fully develop at $\tilde{X}_w \approx 2 \times 10^4$, which correlates well with the set above value $\tilde{\omega}_m = 0.83$ for the fully developed roughness.

A more general approximation form of the gravity wave spectrum covering both developing and fully developed roughness was obtained in the Joint North Sea Wave Project (JONSWAP) experiment (Hasselmann et al. 1973) and is currently known as JONSWAP spectrum:

$$G_z(\omega) = \alpha g^2 \omega^{-5} \exp\left[-1.25\left(\frac{\omega_m}{\omega}\right)^4\right] \exp\left\{\ln\gamma \exp\left[-\frac{(\omega - \omega_m)^2}{2\sigma^2 \omega_m^2}\right]\right\} \qquad (2.25)$$

It can be easily noticed that Eqn. (2.21) is nothing but the Pierson–Moskowitz spectrum multiplied by the peak-enhanced factor, which depending on γ changes the value and the degree of spectral maximum sharpness. Here

$$\sigma = \begin{cases} 0.07, & \omega \leq \omega_m \\ 0.09, & \omega > \omega_m \end{cases}$$

and parameter γ can be obtained from the following table (Davidan et al. 1985):

Table 1   Correlation between $\tilde{\omega}_m$ and γ

$\tilde{\omega}_m$	0.8	1.0	1.3	1.7	2.0
γ	0.9	1.05	2.12	2.48	4.36

The coefficient α can be found taking into account the empirical relationship (Hasselmann et al. 1973):

$$\sigma_z^2 = \frac{1.6 \times 10^{-7} \, \tilde{X}_w U^4}{g^2} \tag{2.26}$$

Since σ_z^2 is almost fully dependent on the gravity windsea, we can equate the right-hand sides of Eqns (2.13) and (2.26):

$$\int_0^\infty d\omega \; G_z(\omega) = \frac{1.6 \times 10^{-7} \, \tilde{X}_w U^4}{g^2} \tag{2.27}$$

where the integrand on the left-hand side of Eqn. (2.27) is defined by Eqn. (2.25). We take the integral with allowance for Eqn. (2.25), and find α as a function of dimensionless windfetch $\tilde{X}_w$. Karaev and Balandina (2000) calculated $\alpha = 8.1 \times 10^{-3}$ for the fully developed windsea $(\tilde{X}_w = 2 \times 20^4)$ (see expression (2.21)) and also showed that the dependence of α on $\tilde{X}_w$ is rather weak – within the interval of transition of $\tilde{X}_w$ from 1.4×10^3 to 2×10^4, i.e. more than by an order, the value of α does not increase more than 1.5 times.

The above-cited formulas together with the table describe completely the JONSWAP model (as well as the Pierson–Moskowitz one as a particular case) of the gravity windsea temporal spectrum.

Speaking about the gravity roughness, we cannot omit its other important parameter, namely, significant wave height

$$H_s \approx 4\sigma_z \tag{2.28}$$

which is the mean of highest one-third of waves observed (note that the wave height is defined as a distance from the wave trough to its crest). A wide use of H_s to characterize roughness is caused by the fact that the H_s value roughly matches the non-instrumental visual assessment of the wave height. (Note that formula (2.28) results from the Gaussian statistics of sea surface elevation.)

With the wind slackening, the roughness spectrum gradually narrows. After the wind completely dies out, the broad spectrum becomes reduced to solely the main peak part. This means the windsea evolves into decaying swell, which can while slightly fading out travel to large distances (hundreds and even thousands of kilometers) from the area where the wind ceases to feed the waves with energy. The following spectrum approximation for this situation has been suggested in Davidan et al. (1985):

$$G_z(\omega) = 6\sigma_z^2 \left(\frac{\omega_m}{\omega}\right)^5 \omega^{-1} \exp\left[-1.2\left(\frac{\omega_m}{\omega}\right)^5\right] \qquad (2.29)$$

If the swell has been generated by the windsea developed at wind speed U, then evidently the value of σ_z in Eqn. (2.29) should be anyway less than that given in Eqn. (2.22).

The issue of the mixed spectrum of windsea plus swell is rather complicated; however, if the respective spectra maximums are sufficiently spaced apart, the combined spectrum can be represented as a sum of windsea and swell spectra.

So far we have dealt with the temporal roughness spectrum. However, the analysis of the radar image of the ocean involves the spatial and not the temporal spectrum and we apparently need to understand the relationship between the two.

As mentioned earlier, there is a dispersion correlation between the frequency and the length of the wave, which appears as in Eqn. (2.20) for gravity roughness. Yet, this correlation deals with wave frequency and length and does not mention anything about its propagation vector. The propagation data are mainly catered by pitch-and-roll oceanographic buoys and stereophotography. Besides, the data are procured by HF radar sensing (see below).

Consider the roughness spatial spectrum $W_z(\kappa, \varphi)$ in the polar coordinates; from Eqn. (2.16), one can write

$$\kappa W_z(\kappa, \varphi) = S(\kappa)n(\kappa)\Phi(\kappa, \varphi) \qquad (2.30)$$

where

$$S(\kappa) = G_z\left[(\kappa g)^{1/2}\right]\left(\frac{d\omega}{d\kappa}\right) = \frac{1}{2}\left(\frac{g}{\kappa}\right)^{1/2}\left[(\kappa g)^{1/2}\right] \qquad (2.31)$$

is the omnidirectional spectrum, φ is the wave direction relative to the wind, $\Phi(\kappa, \varphi)$ is the spreading function and

$$n(\kappa) = \frac{1}{\displaystyle\int_0^{2\pi} \Phi(\kappa, \varphi)\ d\varphi} \qquad (2.32)$$

Thus,

$$\sigma_z^2 = \int_0^\infty S(\kappa)\ d\kappa \qquad (2.33)$$

while for JONSWAP spectrum (see Eqn. (2.25)) the omnidirectional spectrum takes the form

$$S(\kappa) = \frac{\alpha}{2}\kappa^{-3}\exp\left[-1.25\left(\frac{\kappa}{\kappa_{\mathrm{m}}}\right)^{-2}\right]\exp\left\{\ln\gamma\exp\left[-\frac{\left(\sqrt{\kappa/\kappa_{\mathrm{m}}}-1\right)^2}{2\sigma^2}\right]\right\} \qquad (2.34)$$

As to spreading function, it can be set as (Longuet-Higgins et al. 1963):

$$\Phi(\kappa,\varphi) = \cos^{2\mathrm{s}(\kappa)}\frac{\varphi}{2} \qquad (2.35)$$

where the exponent $2\mathrm{s}(\kappa)$ determines the angular width of the spectrum, i.e. $2\mathrm{s}(\kappa)$ is the spreading factor. Apparently, the higher the value of $2\mathrm{s}(\kappa)$, the higher the concentration of the respective roughness spectrum component energy in the wind direction. It is rather difficult to measure $2\mathrm{s}(\kappa)$; however, the estimates can be obtained from experimental data and roughness spectrum models (see e.g. Mitsuyasu et al. 1975, Fung and Lee 1982, Apel 1994, Elfouhaily et al. 1997).

Notably, Eqn. (2.35) yields $\Phi(\kappa,\pm\pi) = 0$; yet experimental data display that the surface is characterized by upwind large-scale gravity waves (although of fairly low intensity) (see below).

2.3 GRAVITY–CAPILLARY WAVE SPECTRA

Though almost all the roughness energy is focused in the gravity part of the spectrum, the gravity–capillary component still plays its specific and no less important role. Apart from initiating roughness development, the gravity–capillary constituent is highly sensitive to various near-surface atmospheric and intra-oceanic processes. Besides, as we will see later, exactly these waves are responsible for electromagnetic scatter within the microwave range which is employed by radar equipment for global monitoring of the ocean.

Figure 2.3 (from Davidan et al. 1985) shows an empirical spectrum of the fully developed roughness at frequencies $\omega \geq 2.5$ rad s^{-1} with wind speed $U = 10$ m s^{-1} (vertical dotted lines indicate gravity–capillary interval limits).

As mentioned earlier, under the realistic conditions high-frequency small-scale waves are influenced by large gravity waves. This influence evokes variations in gravity–capillary waves intensity depending on their position on the large-scale wave profile; in this case as the experimental data show, small-scale ripples are situated for the most part on the front slope of the large-scale wave, next to its crest. This phenomenon is known as hydrodynamic modulation and, as we will see later, is one of the main reasons why microwave radar perceives large gravity waves. The hydrodynamic modulation theory cannot be considered as completely established and thorough; however, we still can single out the main mechanisms responsible for the modulation of the short waves

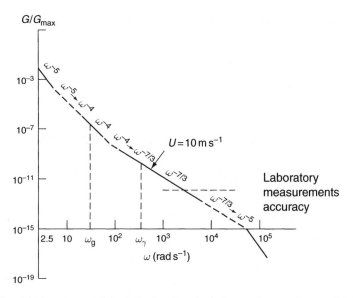

Figure 2.3 Empirical spectrum of the fully developed windsea at frequencies $\omega \geq 2.5$Hz and wind speed $U = 10$ ms^{-1}. Vertical lines indicate gravity–capillary interval limits (Davidan et al. 1985).

spectral density. They are, first, transformation of the short waves spectrum in the non-homogeneous currents area of the long wave and, second, short waves increment modulation induced by different conditions under which they are generated along the large wave profile (Hara and Plant 1994, Thompson and Gotwols 1994, Troitskaya 1994, 1997).

As small-scale ripples participate in the orbital movement within the range of velocities created by the long wave, the dispersion correlation (2.19) stops working here and the transition from the ripples frequency spectrum in Figure 2.3 to the wave number spectrum turns troublesome. For solution to this problem one may apply one of the existing theoretical models of the wave number spectrum, e.g. Elfouhaily et al. (1997). This two-regime model based on the JONSWAP spectrum in the long-wave regime and on the works of Phillips (1985) and Kitaigorodskii (1973) at high wave numbers embraces the whole roughness spectrum from large gravity waves to the smallest capillary ones.

Figure 2.4 (from Thompson 2004) displays omnidirectional wave number spectra corresponding to the model of Elfouhaily et al. (1997) under various wind speed conditions. Noteworthy is the fact that the spectral density is almost independent of the wind speed in the wave number range between about 1 and 10 rad m^{-1}. This is exactly the equilibrium sub-range we have mentioned above where there is a balance between the energy induced by the wind and the one transmitted to the long-wave part of the spectrum.

Figure 2.5 (also from Thompson 2004) shows the diagrams of the exponent $2s(\kappa)$ from Eqn. (2.35). Their characteristic feature is nonmonotone curves $2s(\kappa)$, each of

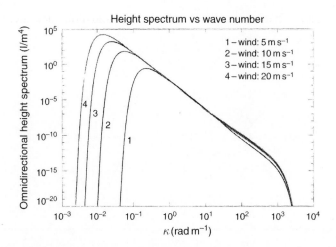

Figure 2.4 Omnidirectional wave number spectra corresponding to the model of Elfouhaily et al. (1997) (cited in Thompson 2004).

which has a maximum in the gravity–capillary part of the spectrum. This maximum apparently can indicate that the respective short waves riding on a long gravity wave point in its propagation direction. Note that these curves have been obtained within the framework of the roughness spectrum theoretical model and need a thorough experimental check, which is fairly difficult under the field conditions.

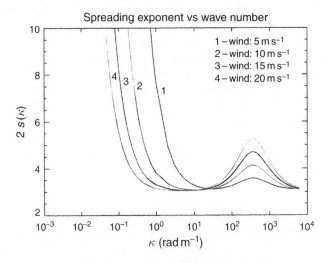

Figure 2.5 Diagrams of the exponent $2s(\kappa)$ from Eqn. (2.35) corresponding to the model of Elfouhaily et al. (1997) (cited in Thompson 2004).

The information on the spreading factor $2s$, quoted from different sources in Kim et al. (2003) is as follows:

$$2s = \begin{cases} \approx 8, & \Lambda = 1.5\,\text{cm} \\ \approx 5, & \Lambda = 5\,\text{cm} \\ \leq 4, & \Lambda > 5\,\text{m} \end{cases}$$

where $\Lambda = 2\pi/\kappa$. According to Lyzenga (1991), originally given in Pierson and Stacy (1973), the spreading factor value varies from 4 to 10 in the transitional interval between the spectral maximum (where $2s$ increases up to 20) and the capillary area.

There is one more feature which will later prove essential in the context of radar remote sensing of the ocean surface. The point is that apart from phase and orbital velocities the gravity–capillary ripples have some constant velocity, even when there is no external surface current (Trizna 1985). The additional constant velocity is a sum of Stokes drift current velocity v_{St} and the speed of wind drift resulting from the continuity of stress across the sea surface as momentum is transferred from wind to waves:

$$v_{wd} = (0.02 - 0.03)U \qquad (2.36)$$

Stokes current drift occurs at rough sea when the trajectories of water particles are not closed and the particles perform cycloidal-type motion with a consequent slow advance in the same direction as the main roughness energy propagates. The speed of the advance motion in deep waters is

$$v_{St} = 2\pi^3 \frac{(H/T)^2}{gT} \qquad (2.37)$$

where H and T are characteristic height and the period, respectively, in the surface roughness. If we assume $v_{orb} \approx \pi H/T$ for the orbital movement velocity, then

$$v_{St} \approx \frac{2\pi}{T} \frac{v_{orb}^2}{g} \approx \omega_m \frac{v_{orb}^2}{g} \qquad (2.38)$$

where ω_m is the frequency of the maximum in the large-scale roughness temporal spectrum. In case of fully developed windsea $\omega_m = 0.83\ g/U$, after we express v_{orb} in terms of its root mean square (RMS) $\sigma_{orb} \approx 0.5\ v_{orb}$ and take into account that $\sigma_{orb} = \sigma_{vert} \approx 6.8 \times 10^{-2}U$ (see Eqn. (2.23)), we obtain the value for Stokes drift current velocity:

$$v_{St} \approx (0.01 - 0.02)U \qquad (2.39)$$

If roughness has the form of swell, v_{St} should be estimated using Eqn. (2.37).

2.4 REALISTIC OCEAN SURFACE AND ITS FEATURES

All that has been said above here refers to "pure" roughness. However ocean roughness (gravity–capillary and short gravity waves mainly) is influenced by various surface, subsurface and atmospheric boundary layer processes. Besides there often are man-made formations – oil spills mostly. Numerous optical and radar ocean graphs prove that the surface is covered all over by inhomogenities of various scale caused by internal waves, eddies, oceanic fronts, surface slicks and so on. Thus, for instance, Figure 2.6 (from Lyzenga and Marmorino 1998), is a radar image of the area that includes the north wall of the Gulf Stream and adjacent shelf near Cape Hatteras; one can also see a number of features, which are explained in the sketch map given together with the image. (Note that this is a fairly rare case where all the features presented within a single frame are identifiable.)

The coastal ocean zone has an especially wide set of spatial and temporal scales (from metres to hundred of kilometres and seconds to several days). Figure 2.7 (from Johannessen 1995) illustrates schematically the spatio-temporal scale of various oceanic processes manifestations in the coastal zone. As can be seen, they mostly lie within the scale characteristic for the surface gravity roughness.

The features shown in Figure 2.6 by far do not exhaust all the features observed on the ocean surface; however, they suffice to make us understand that retrieving comprehensive data on the ocean roughness on the basis of radar imagery is not simple. Therefore, radar imagery interpretation should be taken up with great care, the more so, as the book suggests later, radar imaging mechanisms distort the image of ocean roughness.

Alongside roughness parameters definition, a very important task pragmatically is identification of slicks (smoothed small-scale roughness areas) with imaging radars and mapping of surface and sub-surface currents.

Slicks are for the most part caused by surface-tension-reducing films invoking damping of gravity-capillary waves. Slicks of biogenic origin are now known to occur in all seas. Originating from marine plants and animals, biogenic slicks are an indicator of biological productivity. The producers of biogenic films in the sea are algae and some bacteria, as well as zooplankton and fish (the latter two are comparatively insignificant). A characteristic feature of the biogenic slicks is that the films are only one molecular layer thick (approximately 3 nm).

Non-biogenic slicks caused by mineral or petroleum oil are man-induced or brought about by the seepage of natural oil from the sea bottom. The primary source of oil pollutions in the ocean is oil tanker accidents such as groundings and collisions, mostly in coastal areas. This leads to heavy oil spills and further to ecological catastrophes. Besides, there are large spills near oil terminals and oil platforms. Pollution of the sea surface by mineral oil is a major environmental problem.

The task of mapping surface/sub-surface currents is also of paramount importance. First, because currents spread warmth over the Earth and enrich its atmosphere with water vapor, oxygen and salts. Second, they transport floating matter, in particular, pollutants and thus are of great importance in coastal areas, where considerable damage can be done by surface-borne pollutants and oil. Besides, many types of fish eggs are borne by surface currents, which are therefore of concern to the fisheries industry.

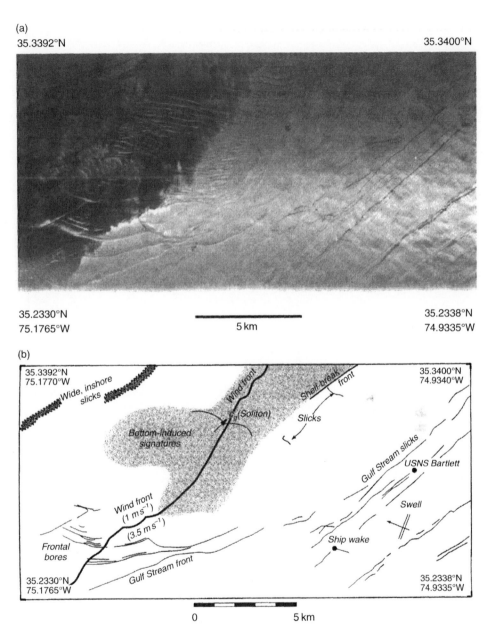

Figure 2.6 Radar image of the ocean surface containing numerous features (a) and the corresponding sketch map (b) (Lyzenga and Marmorino 1998).

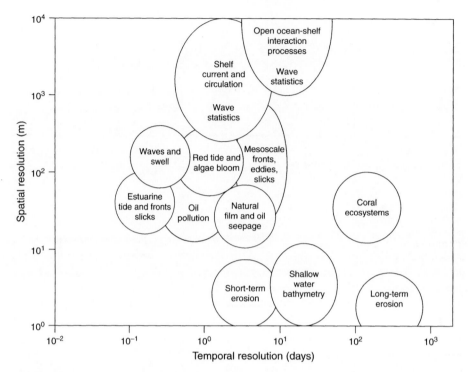

Figure 2.7 Spatio-temporal scale of various oceanic processes manifestations in the coastal zone (Johannessen 1995).

Slicks and currents (as well as other features) can be seen in the radar imagery only owing to their impact on the surface roughness. Therefore, the tasks of estimating roughness parameters and identifying ocean features are simultaneously interlinked.

– 3 –

Sea scattering of radio waves

For quite a long time and particularly for the past three decades since the launch of the first oceanographic satellite SEASAT, there have been a number of radar-equipped satellites on the orbit operating in various frequency intervals and under different observation angles. During this time innumerable data have been compiled relating to different aspects of radar remote sensing of the ocean. At the same time the theory of radar wave scattering from a rough moving surface, which is the milestone of the radar probing results interpretation, cannot be considered as absolutely complete in its current state. Evidently, it would be extremely desirable to formulate a unified theory that might embrace a broader range of survey and environmental conditions. Although considerable effort has been applied here which has brought about important success (Fung 1994, Voronovich 1994, Elfouhaily et al. 1999, Elfouhaily et al. 2001, Elfouhaily and Johnson 2007), we cannot so far speak about a unified scatter theory. Therefore, we will further use different approximations appropriate for the given remote-sensing conditions.

3.1 SEA WATER DIELECTRIC CONSTANT AND ELECTROMAGNETIC PENETRATION DEPTH

A parameter characterizing electric properties of sea water as a substance when interacting with electromagnetic field is known as a dielectric constant

$$\varepsilon = \varepsilon' + i\varepsilon'' \tag{3.1}$$

where ε' is the real part, ε'' is the imaginary part and $i = \sqrt{-1}$. The real part, ε', describes the ability of a medium to store electrical energy, and the imaginary part, ε'', commonly termed the loss factor, describes the electromagnetic loss of the medium. The loss tangent

$$\tan \delta = \frac{\varepsilon'}{\varepsilon''} \tag{3.2}$$

describes whether the material is a good conductor (large loss tangent, $\tan \delta \gg 1$) or poor conductor (low loss tangent, $\tan \delta \ll 1$).

Let the electromagnetic wave propagate in a medium with a dielectric constant ε, i.e.

$$w \propto \exp\left[i\left(k\sqrt{\varepsilon}x - \omega t\right)\right] \tag{3.3}$$

where $k = 2\pi/\lambda$ is the electromagnetic wave number. We set $\sqrt{\varepsilon}$ as

$$\sqrt{\varepsilon} = n + i\chi \tag{3.4}$$

where n is the index refraction, and $\chi = \alpha/k$ with α being the attenuation coefficient:

$$n = \sqrt{|\varepsilon|}\cos\left(\frac{1}{2}\tan^{-1}\delta\right), \quad \chi = \sqrt{|\varepsilon|}\sin\left(\frac{1}{2}\tan^{-1}\delta\right) \tag{3.5}$$

Then

$$w \propto \exp(-\alpha x)\exp[i(knx - \omega t)] \tag{3.6}$$

i.e. a wave propagating in medium with a complex dielectric constant fades out sharing its energy with the medium. The distance where the energy decreases by e times, or in fact the penetration depth of the electromagnetic field into medium, is

$$\delta_{\mathrm{p}} = \frac{\lambda}{4\pi\sqrt{|\varepsilon|}\sin[(1/2)\tan^{-1}\delta]} \tag{3.7}$$

The real and imaginary parts of dielectric constant of fresh and sea water, as well as penetration depth are represented in Figures 3.1 and 3.2. One can see that in the microwave range the penetration depth of the electromagnetic field under sea surface does not exceed fractions of a centimetre.

As for reflectivity, for incidence of a plane electromagnetic wave onto plane sea surface the reflection coefficients (the ratio of the reflected field amplitude to the incident field amplitude) are given by Fresnel's well-known formulas

$$F_{\mathrm{V}} = \frac{\varepsilon\cos\theta_0 - \sqrt{\varepsilon - \sin^2\theta_0}}{\varepsilon\cos\theta_0 + \sqrt{\varepsilon - \sin^2\theta_0}} \tag{3.8a}$$

$$F_{\mathrm{H}} = \frac{\cos\theta_0 - \sqrt{\varepsilon - \sin^2\theta_0}}{\cos\theta_0 + \sqrt{\varepsilon - \sin^2\theta_0}} \tag{3.8b}$$

where subscripts V and H denote vertical and horizontal polarizations of the incident wave, respectively; and θ_0 the incidence angle. (Remember that a vertical polarization wave is the one with electric field vector in the incidence plane; in case of horizontal polarization this vector is perpendicular to the incidence plane and parallel to the two media divide.) Figure 3.3 shows dependencies $F_{\mathrm{V}}(\theta_0)$ and $F_{\mathrm{H}}(\theta_0)$ for the electromagnetic wave $\lambda = 3\,\mathrm{cm}$. Note the minimum of the curve $F_{\mathrm{V}}(\theta_0)$ at $\theta_0 \approx 83°$. The incidence angle of

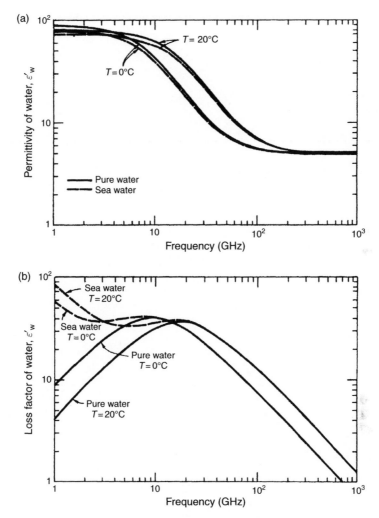

Figure 3.1 (a) Real part of the complex dielectric constant of pure water and sea water. (b) Imaginary part of the complex dielectric constant of pure water and sea water. Salinity of sea water is 32.45‰ (salinity units) (Holt 2004; originally cited from Ulaby et al. (1986)).

the minimum depends on the dielectric constant ε; it is termed *Brewster's angle*. Brewster's angle changes the nearby reflection coefficients for the two polarizations – this is a feature peculiar to the microwave scatter at low grazing angles (see Section 3.3).

As we see, microwaves are generally rather well reflected by the sea surface and do not almost penetrate into the water. This means that the intra-oceanic processes manifest themselves in the microwave radar image of the ocean through surface roughness changes incited by deep subsurface processes, rather than overtly (of course, on the condition these changes are significant enough.)

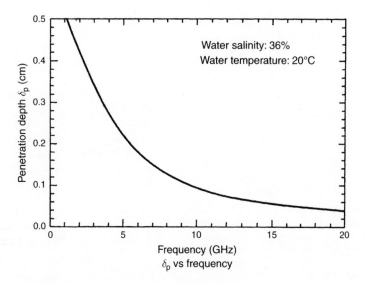

Figure 3.2 Penetration depth of sea water as a function of frequency (Holt 2004; originally cited from Ulaby et al. (1986)).

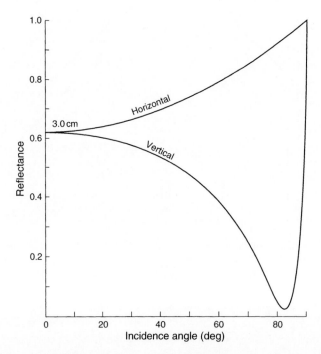

Figure 3.3 Fresnel coefficients (reflectance) of a plane sea-water surface as a function of incidence angle for the radiation wavelength $\lambda = 3$ cm. "Horizontal" and "vertical" refer to the polarization of the radiation (from Stewart 1985).

3.2 RESONANT SCATTERING

This section is dedicated to the resonant mechanism of scattering of radio waves by the rough ocean surface. The mechanism, in one form or another, works both at decameter radio waves (HF diapason) and microwaves, and exactly in the area of incident angles which radar imaging of the ocean usually works with. Therefore, most experimental results obtained in this scope are interpreted on the basis of resonant scattering.

3.2.1 HF scattering and HF radars

The scattering of radio waves by the sea (ocean) surface has been studied since the Second World War. However, the physics of the phenomenon was first explained as early as mid-1950s using the data from HF coast radar experiments [Crombie 1955, Braude 1962]. As mentioned in the Introduction, those were the first experiments to regard the sea as the study object rather than the source of perturbations and interferences.

A typical Doppler spectrum of the backscattered HF signal (frequency $f_0 = 13.4$ MHz) is shown in Figure 3.4 (quoted from Barrick (1978)). This spectrum was yielded by an

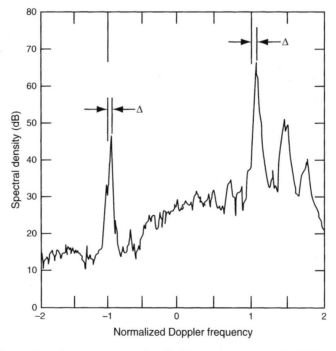

Figure 3.4 Measured surface-wave sea-echo Doppler spectrum at 13.4 MHz. The Doppler frequency axis is normalized, with 0 corresponding to the transmitter carrier frequency position, and ± 1 being the first-order Bragg frequency. A small shift $\Delta = (\Delta f_0 - F_{res})/F_{res}$, where Δf_0 is the Doppler shift observed in the experiment, points to local current off California coast of the United States (from Barrick 1978).

experiment carried out on San Clemente Island (CA, USA). The scattering patch was located 30 km westward from the radar, and was approximately 3 km in radial extent by 5 km in azimuthal extent. The spectral resolution in this plot is about 0.005 Hz.

The main spectrum feature is the two well-defined lines spaced apart against probing frequency; their intensities differ considerably (usually by 15–20 dB). Notably, at a given frequency f_0 the line positions are not dependent on the sea surface state. On probing the frequency changeover, the spectrum lines shift proportionally by $f_0^{1/2}$, but not f_0, as could be supposed in view of Doppler shift on fixed scatterers.

These facts are explained by microwave rough surface scattering theory represented by perturbation theory. It applies well for the HF waves:

$$\frac{\sigma_z^2}{\lambda^2} \ll 1, \quad \left\langle (\nabla z)^2 \right\rangle \ll 1 \tag{3.9}$$

where σ_z is the RMS deviation of the scattering surface $z(x, y)$ from the averaging surface $z = 0$ and λ stands for radar wavelength (angle brackets represent averaging over surface realizations). Evidently, these conditions imply that the roughness has to be slightly sloping as well as small against λ.

According to the first (Bragg's) approximation of the perturbation theory, the radar cross section (see the definition in Chapter 1) of the surface area S_0 illuminated along the direction x under depression angle ψ_0 is represented by the following expression (Wright 1968):

$$\sigma = \frac{4k^4}{\pi} \left| \hat{g}(\varepsilon, \psi_0) \right|^2 \left\langle \left| \iint\limits_{S_0} dx\, dy \cdot z(x, y) \exp(2ikx \cos \psi_0) \right|^2 \right\rangle \tag{3.10}$$

Here $k = 2\pi/\lambda$, ε is the dielectric constant of the lower medium (here, sea water), and the function $\hat{g}(\varepsilon, \psi_0)$, dependent on the radiated field polarization, is given by (Valenzuela 1978)

$$\hat{g}(\varepsilon, \psi_0) = \begin{cases} g_{HH} \sin^2 \psi_0, & \text{horizontal polarization} \\ g_{VV} \sin^2 \psi_0, & \text{vertical polarization} \end{cases} \tag{3.11}$$

$$g_{HH} = \frac{(\varepsilon - 1)}{\left[\sin \psi_0 + (\varepsilon - \cos^2 \psi_0)^{1/2} \right]^2} \tag{3.12a}$$

$$g_{VV} = \frac{(\varepsilon - 1)[\varepsilon(1 + \cos^2 \psi_0) - \cos^2 \psi_0]}{\left[\varepsilon \sin \psi_0 + (\varepsilon - \cos^2 \psi_0)^{1/2} \right]^2} \tag{3.12b}$$

Here the transmitted and received electromagnetic fields are co-polarized.

Provided the linear dimensions of the irradiated patch drastically exceed the respective scale of elevations correlation (i.e. spatial scale marked by significant receding of correlation function $B_z(\vec{\rho}, 0)$), with the help of expressions (3.2)–(3.4), we find

$$\left\langle \left| \iint\limits_{S_0} dx\, dy\, z(x,y) \exp(2ikx \cos \psi_0) \right|^2 \right\rangle = 4\pi^2 S_0 \hat{W}_z(-2k \cos \psi_0, 0) \qquad (3.13)$$

where $\hat{W}_z(\kappa_x, \kappa_y)$ is a spatial spectrum of the surface elevations against plane $z = 0$. Thus the NRCS $\sigma_0 = \sigma/S$ is defined by

$$\sigma_0 = 16\pi k^4 |\hat{g}(\varepsilon, \psi_0)|^2 \hat{W}_z(-2k \cos \psi_0, 0) \qquad (3.14)$$

This displays the resonance character of the scatter, for σ_0 is defined as the only component of the spectrum $\hat{W}_z(\vec{\kappa})$; here, in case of backscattering, it is a component with resonance wave number $\vec{\kappa}_{res} = \{-2k \cos \psi_0, 0\}$.

Central symmetry (that is to say, symmetry against the origin) of the spatial spectrum is produced by two, in general case, overlapping "halves", $\hat{W}_z^{\pm}(\vec{\kappa})$ and

$$\hat{W}_z^+(\vec{\kappa}) = \hat{W}_z^-(-\vec{\kappa}), \quad \hat{W}_z^+(-\vec{\kappa}) = \hat{W}_z^-(\vec{\kappa}) \qquad (3.15)$$

(the "+" or "−" sign applies to the frequency ω in the temporal factor $\exp(-i\omega t)$ of the propagating wave). Therefore

$$\hat{W}_z(\vec{\kappa}_{res}) = \hat{W}_z^+(\vec{\kappa}_{res}) + \hat{W}_z^-(\vec{\kappa}_{res}) = \hat{W}_z^+(\vec{\kappa}_{res}) + \hat{W}_z^+(-\vec{\kappa}_{res}) \qquad (3.16)$$

Remember (see Eqn. (2.15)) that after we restrict ourselves to positive frequencies ω, we get the following expression:

$$\hat{W}_z^+(\vec{\kappa}) = \frac{1}{2} W_z(\vec{\kappa}) \qquad (3.17)$$

and consequently the formula for NRCS can be represented as

$$\sigma_0 = 8\pi \kappa^4 |\hat{g}(\varepsilon, \psi_0)|^2 \left[W_z\left(+\vec{\kappa}_{res}\right) + W_z\left(-\vec{\kappa}_{res}\right) \right] \qquad (3.18)$$

where the pair $\pm\vec{\kappa}_{res}$ corresponds to two counter-propagating surface waves, i.e. directed towards and from the radar.

Expression (3.18) can be then rewritten as

$$\sigma_0 = \sigma_0^+ + \sigma_0^- \qquad (3.19)$$

and thus the ratio σ_0^+/σ_0^- is nothing but an intensity ratio of approaching and receding waves.

At small depression angles ψ_0 characteristic for coastal HF radars,

$$\Lambda_{\text{res}} = \frac{2\pi}{\kappa_{\text{res}}} \approx \frac{\lambda}{2} \tag{3.20}$$

i.e. the scatter source is two counter-propagating meter or decameter waves with the following frequency according to the dispersion relation for the gravity part of the spectrum:

$$F_{\text{res}} = \frac{1}{2\pi}\left(\kappa_{\text{res}}g\right)^{1/2} \propto f_0^{1/2} \tag{3.21}$$

where $g = 9.8\,\text{m s}^{-2}$ is the gravity acceleration and the phase velocity is

$$v_{\text{ph.res}} = F_{\text{res}}\Lambda_{\text{res}} \tag{3.22}$$

which provides the shift of the backscattered signal frequency:

$$\Delta f_0 = \frac{2v_{\text{ph.res}}}{\lambda}\cos\psi_0 = F_{\text{res}} \tag{3.23}$$

Therefore the Doppler spectrum has two lines with frequencies $\pm F_{\text{res}} \propto f_0^{1/2}$, and the intensities are each proportional to the intensity of the corresponding wave; these lines are represented in Figure 3.4.

The fact that both lines are conspicuous in the real Doppler spectrum testifies to the existence of upwind waves (though of low intensity) on the ocean surface alongside downwind waves. As shown in Crombie et al. (1978), these waves are caused by weak non-linear interaction within the wave spectrum.

One more feature of the Doppler spectrum is a broad echo continuum surrounding the lines. This continuum is not a radar system noise, but varies both in amplitude and shape with sea state and radar frequency, which is supported by numerous experiments. As shown in Barrick (1978), the continuous part of the spectrum (in low HF band at roughly $-30\,\text{dB}$ from the line peaks) is well described by the second approximation of the perturbation theory. The second-order echo gains significance in the HF part of the HF range starting approximately at 20 MHz.

Now let us consider the potentialities of HF radar in the remote sensing of the ocean. First, note that HF range waves travel very far. They have two alternative modes: "ground wave" and "sky wave". The former concerns the diffraction by the curved earth, which allows the vertically polarized HF field to propagate above the well-conducting water surface over distances much further than the horizon. As Barrick (1978) notes, radars located at ocean level - operating by "ground wave" mode - can observe sea echo as far away as 200 km. The second mode ("sky wave") is associated with ionosphere acting as a

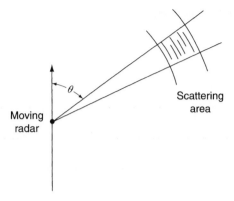

Figure 3.5 Plan view of geometry for viewing a scattering area using an antenna synthesized by moving HF radar (from Stewart 1985).

concentric mirror at 100–400 km above the Earth for radar wave frequencies lower than 30 MHz. However, this "mirror" is not steady; as Stewart (1985) expresses it, it behaves as a wavy, distorting, absorbing, moving reflector. Nevertheless, by carefully choosing the time and frequency, the accurate observations of radio scatter can be made over the range 1000–4000 km.

However, alongside the benefits mentioned above, there is an intrinsic significant drawback of the HF radar – the complicated formation of a sufficiently directed radar beam requiring extended (over 100 m long) phased-array antenna systems. Nevertheless, such antennas are in use and employed for mapping of ocean surface features (see below) electronically steering a narrow beam and scanning over extensive coastal areas.

Besides, there is another widely used and less expensive method to achieve high angular resolution as described in Stewart (1985).

Let the radar with a wide antenna pattern move linearly at the speed v (Figure 3.5). Doppler frequency of the radiation returned at an angle of ϑ is

$$f = \frac{2v}{\lambda} \cos \vartheta \tag{3.24}$$

where λ is the radar wavelength. It is evident that the Doppler frequency resolution determines angle resolution:

$$df = -\frac{2v}{\lambda} \sin\vartheta \; d\vartheta \tag{3.25}$$

In vicinity of $\vartheta = \pi/2$

$$|df| \approx \frac{2v}{\lambda}|d\vartheta| \tag{3.26}$$

The Doppler frequency resolution is inversely proportional to the duration T of the sample used to compute the Doppler spectrum:

$$\mathrm{d}f \approx \frac{1}{T} \tag{3.27}$$

Within this time spell, the radar travels across the distance of $L = vT$, thus

$$\mathrm{d}\vartheta \approx \frac{\lambda}{2L} \tag{3.28}$$

One can see that the method described provides angular resolution, yielded by a conventional linear antenna of $2L$ length. This method works in the lower HF band. The reason for this restriction is that the increase in frequency brings about the expansion of Bragg lines, and Doppler shifts caused by radar movement can entail the overlap of the lines corresponding to approaching and receding waves.

Figure 3.6 presents the example of ocean wave directional spectra obtained with this method. Another important feature of HF radar is its ability to measure the radial component of the ocean surface current averaged to a depth on the order of one-eighth of the wavelength of the resonant ocean wave. If an ocean wave is carried along by a surface current, its radial velocity differs from the theoretical one by the radial component of the current velocity. It means that the Doppler spectrum lines will be shifted against $\pm F_{res}$ in either direction depending on the current. (In particular, in Figure 3.4 there is a small shift $\Delta = (\Delta f_0 - F_{res})/F_{res}$, where Δf_0 is the Doppler shift observed in the experiment, pointing to a local current off the California coast in the United States.)

Mapping surface and sub-surface currents is of highest importance, as these currents transport floating matter and thus are of great importance in coastal areas, where considerable damage can be done by surface-borne pollutants and oil. This task alongside microwave radar tools (see below) is accomplished in the coastal regions with the help of HF radars. Thus, Barrick et al. (1977) and Barrick (1978) give mapping of currents at a grid of points $3 \times 3\,\mathrm{km}^2$, covering areas exceeding $2000\,\mathrm{km}^2$, out to a distance of about $70\,\mathrm{km}$ from the shore. A recent study (Shay et al. 2007) reports on HF radar mapping of surface currents at west Florida shelf over $40 \times 80\,\mathrm{km}^2$ with a $1.2\,\mathrm{km}$ horizontal resolution. Comparison to sub-surface measurements from moored tools revealed RMS differences of $1–5\,\mathrm{cm\ s}^{-1}$, although in some cases the discrepancies are fairly large. Ohlmann et al. [2007] analyse the accuracy of estimating the velocity of currents and the possible differences between HF radar data and in situ the information obtained with a grid of drifters.

The depth over which the current is measured depends on the wavelength of the ocean waves. Short waves are carried along only by thin surface currents, while longer waves are carried by thicker layers. The thickness is determined by the depth of appreciable wave motion, and is about one-eighth of the wavelength of the ocean wave scattering the radio wave (Stewart 1985).

The dispersion relation of surface gravity waves propagating on a horizontally uniform current with a vertical shear can be presented in the form (Ivonin et al. 2004)

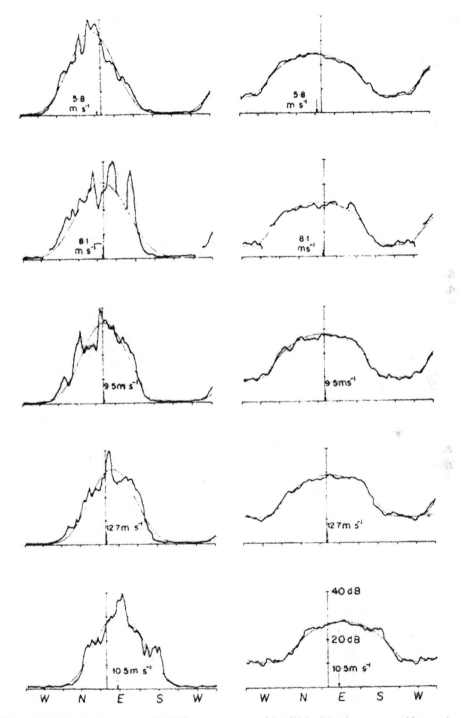

Figure 3.6 Directional spectra of 0.14 Hz waves approaching Wake Island as measured by moving HF radar at 1.95 MHz (Tyler et al. 1974). The energy density on a linear scale (left) and logarithmic scale (right) is plotted; smooth curves are least-squares fits. Wind averages over preceding 8 h are indicated (from Barrick 1978).

$$\omega = \sqrt{g\kappa} + \kappa u_{\text{eff}}(\kappa) \cos \beta \tag{3.29}$$

Here κ is the wave number of the ocean wave, β is the angle between the waves and the mean current and u_{eff} is the effective current velocity, which can stand for the above-mentioned velocity averaged over the depth of about one-eighth of the wavelength. The effective current velocity is well approximated by the formula (Stewart and Joy 1974):

$$u_{\text{eff}}(\kappa) = 2\kappa \int\limits_{-\infty}^{0} u(z) e^{2\kappa z} \, dz \tag{3.30}$$

Here $u(z)$ is the vertically sheared mean current, z is the depth and κ is the wave number of the ocean wave.

The feasibility of current vertical shear appreciation with a multi-frequency radar system by position of first-order peaks in the Doppler spectrum is clear from the theoretical point of view and is substantiated by experiments (see Teague et al. 1977, Stewart 1985). Moreover, the experimental results (Shrira et al. 2001, Ivonin et al. 2004) prove that the vertical shear of sub-surface velocity can be determined by commonly used single-frequency HF radar, if accompanied by two second-order peaks described by the second approximation of the perturbation theory alongside the first-order Bragg peaks.

As mentioned above it is clear that the HF radar is an effective tool for the remote sensing of the ocean (mainly in the coastal areas).

3.2.2 Microwave scattering: two-scale model of the sea surface

With transition to microwaves, one would assume, the scattering mechanism should drastically change as one of the principal conditions of the perturbation theory is violated – the roughness height cannot be regarded as small against the microwave length. However, it turned out that the scatter resonance character is retained in microwave diapason, though modified.

When the angle of incidence of, e.g. centimetre microwaves is not very steep the resonance wave number $\kappa_{\text{res}} = 2k \cos \psi_0$ is also within the centimetre (i.e. gravity–capillary) area of the small-amplitude wave spectrum. Moreover, for the given λ there is always a wave number such that $\kappa_1 < \kappa_{\text{res}}$ and that the small-scale wave RMS height inferred via integration over the area $\kappa > \kappa_1$ is small compared to λ, i.e. the perturbation theory is applicable for this part of the spectrum. In fact, experimental data show that with no large waves and surface currents, backscattered microwave signal frequency shift is defined by the dispersion relation, as the perturbation theory suggests, as of now for the gravity capillary part of the spectrum:

$$\Delta f_0 = F_{\text{res}} = \frac{1}{2\pi} \left(\kappa_{\text{res}} g + \kappa_{\text{res}}^3 \frac{\alpha_s}{\rho_w} \right)^{1/2} \tag{3.31}$$

where α_s is the surface tension, ρ_w the water density and g the acceleration due to gravity; $\alpha_s/\rho_w = 74\,\mathrm{cm}^3\,\mathrm{s}^{-2}$ for a clean water surface.

Substituting $\kappa_{\mathrm{res}} = (4\pi/\lambda)\cos\psi_0$ into Eqn. (3.31), we obtain

$$\Delta f_0 = \left(\frac{g\cos\psi_0}{\pi\lambda} + \frac{16\pi\alpha_s\cos^3\psi_0}{\rho_w\lambda^3} \right)^{1/2} \tag{3.32}$$

Figure 3.7 (quoted from Rozenberg et al. 1966) shows the curve with Doppler shift experimental values plotted for various λ within broad diapason (for microwaves at weak wind and in the absence of large-scale waves and surface currents); it expressly displays an agreement among HF, VHF and microwave scatter. One can see that the experimental points dislocate on the curve when calculated on the basis of Eqn. (3.32), but not the dotted lines corresponding to the Doppler shift for some fixed velocity v.

What happens when the length of large-scale sea waves is many times the microwave length? We shall divide arbitrarily full ocean spectrum into small- and large-scale parts and introduce for the purpose a borderline wave number κ_b. The small-scale part $\kappa > \kappa_b$ is held obeys the perturbation theory, while the large-scale part requires conformity with physical optics (Brekhovskikh 1952):

$$kR\sin^3\psi_0 \gg 1 \tag{3.33}$$

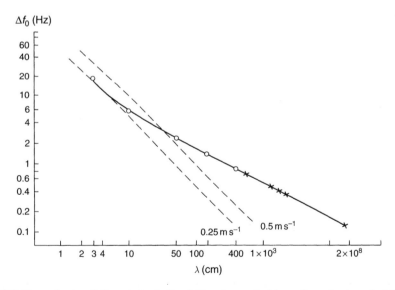

Figure 3.7 Dependence of the returned signal Doppler centroid on the radar wavelength (from Rozenberg et al. 1966). The dotted lines show the calculation results based on the formula $\Delta f_0 = (2v/\lambda)\cos\psi_0$ for the two values of velocity v. One can see that the experimental points dislocate on the curve calculated on the basis of Eqn. (3.32), but not the dotted lines corresponding to the Doppler shift for some fixed velocity v.

R being a mean curvature radius. The latter inequation means that for electromagnetic diffraction, the respective large-scale surface $\zeta(\vec{r}, t)$ can be replaced at any of its points by a tangent plane with a local normal.

As mentioned above, the small-scale part contributes to backscattered signal only by resonance component κ_{res} (more exactly, a narrow area near κ_{res}, whose width is dependent on scattering patch size). Therefore, in terms of scattering, an equivalent surface model is as follows:

$$z(\vec{r}, t) = \zeta(\vec{r}, t) + \xi(\vec{r}, t) \tag{3.34}$$

where $\xi(\vec{r}, t)$ stands for resonance ripples with a narrow spatial spectrum concentrated near κ_{res}, which ride on the large wave $\xi(\vec{r}, t)$. The so-called "two-scale model" (3.34) described in detail in Bass et al. (1968) and Wright (1968) facilitates essentially electro-magnetic microwave field scattering by the surface, which can now be replaced by rough tangent planes at any point.

An additional quantitative justification of the two-scale model can be found in Lementa (1980), where the mean curvature radius

$$R = \left[\int_{0}^{\kappa_2} \kappa^4 W_\varsigma(\vec{\kappa}) \, d\vec{\kappa} \right]^{-1/2} \tag{3.35}$$

for the large-scale surface spectrum $W_\varsigma(\vec{\kappa})$ has been calculated, based on the Pierson–Moskowitz model (Pierson and Moskovitz 1964). These calculations have shown that for each microwave λ, there exists an interval $\kappa_1 < \kappa < \kappa_2$, which obeys both perturbation theory and physical optics method. This allows us to divide the roughness spectrum, and the borderline κ_b should be within the interval. As shown in Lementa (1980), the shift of κ_b within the stated limits does not affect significantly the characteristics of the reflected field, which in fact makes the two-scale model work. The immediate calculations (Lementa 1980) have been performed in the temporal frequency ω domain, as wave spectra, including the Pierson–Moscowitz one, are initially determined in the frequency domain, and their transition into the spatial spectra is effected through the dispersion relation. (Roughness spectrum division has also been discussed in Pereslegin (1975a, 1975b).)

Thus, microwave diapason retains the resonance scatter mechanism which becomes localized, implying that the electromagnetic wave depression/incidence angle is defined against a particular element of the large-scale surface with the localized spectral density of the ripples, non-uniform along the large wave, but not against the horizontal plane.

Formula (3.14) shall now apply to the quasi-flat element of the surface $\zeta(\vec{r}, t)$. It demonstrates that large-scale waves modulate local normalized cross section of the scatter $\sigma_0(\vec{r}, t)$ due to, first, changes of local incidence angle and, second, non-uniformity of the ripples spectral density along the surface ζ.

Modulation caused by local depression angle variations is of purely geometric nature and is therefore called geometrical (or tilt) modulation. Local depression angle depends on the surface $\zeta(\vec{r}, t)$ slopes in the incidence plane and the one perpendicular to it. The dependence of the local cross section on quasi-flat area two-plane slopes (see Valenzuela 1978, Alpers et al. 1981) is given by

$$\sigma_{0,\text{HH}} = 4\pi k^4 \sin^4 \psi_0 \left| \left(\frac{\cos(\psi_0 + \alpha_x) \cos \alpha_y}{\cos \psi_0} \right)^2 g_{\text{HH}} + \left(\frac{\sin \alpha_y}{\cos \psi_0} \right)^2 g_{\text{VV}} \right|^2$$
$$\times \hat{W}_\xi \left(-2k \cos(\psi_0 + \alpha_x), -2k \sin(\psi_0 + \alpha_x) \sin \alpha_y \right) \qquad (3.36a)$$

$$\sigma_{0,\text{VV}} = 4\pi k^4 \sin^4 \psi_0 \left| \left(\frac{\cos(\psi_0 + \alpha_x) \cos \alpha_y}{\cos \psi_0} \right)^2 g_{\text{VV}} + \left(\frac{\sin \alpha_y}{\cos \psi_0} \right)^2 g_{\text{HH}} \right|^2$$
$$\times \hat{W}_\xi \left(-2k \cos(\psi_0 + \alpha_x), -2k \sin(\psi_0 + \alpha_x) \sin \alpha_y \right) \qquad (3.36b)$$

where $\alpha_x = \tan^{-1}(\partial\zeta/\partial x)$ and $\alpha_y = \tan^{-1}(\partial\zeta/\partial y)$ are the slopes of surface ζ in the plane of incidence and in a plane perpendicular to the plane of incidence, respectively; g_{HH} and g_{VV} are determined by Eqn. (3.12). Contrary to Valenzuela (1978), in formulas (3.36) of arguments $\hat{W}_\xi$, besides the grazing angle instead of the incidence one, a "minus" sign also occurs – this difference is not crucial just like the sign is not critical for finding the resonance wave number, i.e. choosing between $\vec{\kappa}_{\text{res}} = \{-2k \cos \psi_0, 0\}$ and $\vec{\kappa}_{\text{res}} = \{2k \cos \psi_0, 0\}$. Notably, Eqn. (3.36) includes the surface slopes and not the local normal to it as in Valenzuela (1978).

Then the local cross section should be averaged over the slopes:

$$\langle \sigma_0 \rangle = \int\limits_{-\infty}^{\infty} d(\tan \alpha_x) \int\limits_{-\infty}^{\infty} d(\tan \alpha_y) \sigma_0 p(\tan \alpha_x, \tan \alpha_y) \qquad (3.37)$$

where $p(\tan \alpha_x, \tan \alpha_y) = p(\partial\zeta/\partial x, \partial\zeta/\partial y)$ is the joint probability density of slopes for the large-scale roughness of the ocean.

Figure 3.8 from Valenzuela (1978) represents the results of the cross section calculation for two polarizations (continuous curves), performed with formula (3.37) at the ripple spectrum $W_\xi \propto \kappa^{-4}$. Besides, the experimental points have been plotted deduced for the wind speed range 11–24 m s^{-1}. Dotted curves apply to small incidence angle area, where quasi-specular reflection mechanisms are at work. The scatter in this area is formed by the reflection from the surface elements perpendicular to the beam and is not resonance (quasi-specular reflection mechanism is explained in detail in Section 3.3).

Modulation of ripples spectral density along the large wave brings about the hydrodynamic modulation. As mentioned in Chapter 1, the main mechanisms of this type of modulation are: first, short-wave spectrum transformation in the inhomogeneous field of long-wave currents and, second, short-wave increment modulation induced by the difference in their formation along the large wave. On this basis, it seems natural to link cross section spatio-temporal modulation to large wave spectrum.

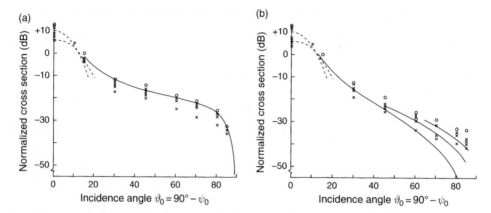

Figure 3.8 Comparison of measured and theoretical cross sections of the ocean surface for a radar frequency of 4455 MHz [Valenzuela 1978]: (a) vertical polarization and (b) horizontal polarization. Solid lines and dotted lines correspond, respectively, to calculations for resonant and quasi-specular (see Section. 3.3) scattering at different surface states. Experimental values are obtained at the wind speed $11 - 24$ m s^{-1} (see Valenzuela [1978] for details).

Suppose that variables σ_0 and ζ are in linear relation:

$$\sigma_0(\vec{r}, t) = \langle \sigma_0 \rangle + \delta\sigma_0 = \langle \sigma_0 \rangle \left[1 + \int d\vec{r}\,' f(\vec{r}\,' - \vec{r}, t) \zeta(\vec{r}\,', t) \right] \qquad (3.38)$$

where $\langle \sigma_0 \rangle$ and $\delta\sigma_0$ represent mean and fluctuation components of the normalized cross section. Yet, the supposition is rather strong and appears not to work always. Particularly, it fails at small depression angles, where according to formulas (3.11) and (3.12), the dependence of σ_0 on the large-scale roughness $\zeta(\vec{r}, t)$ slopes and, consequently, on its elevations is essentially non-linear. However, at low grazing angles, apart from resonance mechanism there are other scatter mechanisms involved (see Section 3.4).

Formula (3.38) yields

$$\sigma_0(\vec{r}, t) = \langle \sigma_0 \rangle \left\{ 1 + \left[\int d\vec{\kappa}\, T(\vec{\kappa}) A_\zeta(\vec{\kappa}) \exp(i(\vec{\kappa}\vec{r} - \Omega t)) + \text{c.c.} \right] \right\} \qquad (3.39)$$

where is A_ζ spectral amplitude in decomposition

$$\zeta(\vec{r}, t) = \int d\vec{\kappa}\, A_\zeta(\vec{\kappa}) \exp\left[i(\vec{\kappa}\vec{r} - \Omega t) \right] + \text{c.c.} \qquad (3.40)$$

Here Ω and κ are in the dispersion relation for the roughness gravity part and "c.c." represents the complex-conjugation variable.

Note that unlike Eqn. (2.7), the surface $\zeta(\vec{r}, t)$ is integrated singularly over spatial wave numbers, and the condition (2.8) bringing about the real character of Eqn. (2.7) does not work here. For this reason in Eqns (3.39) and (3.40), we add the complex conjugate values to the integrals.

The complex function $T(\vec{\kappa})$ is called modulation transfer function (MTF) and is usually written as a sum (Alpers et al. 1981):

$$T(\vec{\kappa}) = T_{\text{tilt}}(\vec{\kappa}) + T_{\text{hydr}}(\vec{\kappa}) \tag{3.41}$$

where T_{tilt} and T_{hydr} are geometric (or tilt) and hydrodynamic parts, respectively. In turn, the geometric part is given by

$$T_{\text{tilt}} = T_{\text{tilt}\parallel} + T_{\text{tilt}\perp} \tag{3.42}$$

$$T_{\text{tilt}\parallel} = \frac{1}{\langle \sigma_0 \rangle} \frac{\partial \sigma_0}{\partial \alpha_x} \cdot i\kappa_x, \quad T_{\text{tilt}\perp} = \frac{1}{\langle \sigma_0 \rangle} \frac{\partial \sigma_0}{\partial \alpha_y} \cdot i\kappa_y$$

where κ_x and κ_y are the components of the wave vector in the plane Z. Such a representation of T_{tilt} is quite comprehensible, as at small slopes

$$\frac{\delta \sigma_0}{\langle \sigma_0 \rangle} \approx \frac{1}{\langle \sigma_0 \rangle} \left(\frac{\partial \sigma_0}{\partial \alpha_x} \alpha_x + \frac{\partial \sigma_0}{\partial \alpha_7} \alpha_y \right) \tag{3.43}$$

and besides

$$\alpha_x \approx \frac{\partial \zeta}{\partial x} = i \int d\vec{\kappa}\, \kappa_x A_\zeta(\vec{\kappa}) \exp\left[i(\vec{\kappa}\vec{r} - \Omega t) + \text{c.c.}\right] \tag{3.44a}$$

$$\alpha_y \approx \frac{\partial \zeta}{\partial y} = i \int d\vec{\kappa}\, \kappa_y A_\zeta(\vec{\kappa}) \exp\left[i(\vec{\kappa}\vec{r} - \Omega t)\right] + \text{c.c.} \tag{3.44b}$$

The geometric part of MTF has been calculated in Alpers et al. (1981) for two wavelengths of microwave diapason at the ripple spectrum $W_\xi(\kappa) \propto \kappa^{-4}$. The calculation results for the wave $\lambda = 23$ cm, such as that on SEASAT, and $\lambda = 3$ cm can be found in Figure 3.9a and b. The upper two curves apply for scattering at range travelling waves and the lower curves for scattering at azimuthally travelling waves. It is apparent that for the slopes of the scattering surface in the plane perpendicular to the incidence, one can hardly modulate the reflected field, as its impact on the local incidence (or grazing) angle is very scarce. As for the dependence on the slopes in the incidence plane, the corresponding curves (Figure 3.9) show weak and strong modulation areas as well as the differences between the fields with opposite polarizations much more clearly than in Figure 3.10.

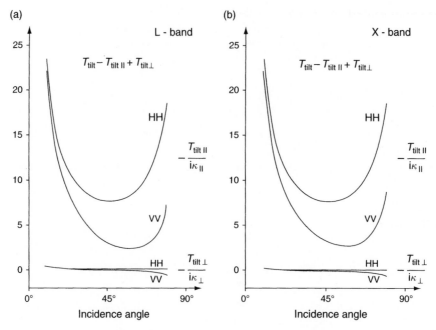

Figure 3.9 (a, b) Dimensionless modulation transfer functions due to tilting of short waves by a long wave as a function of incidence angle, with polarization as a parameter. $R_\parallel$ and $R_\perp$ refer to ocean waves whose crests are parallel and perpendicular to the radar velocity vector. The curves apply to a 1.2 GHz radar, such as that on SEASAT, and 10 GHz radar [Alpers et al. 1981].

At incidence angles $\theta_0 \leq 60°$ (still remaining within the limits of resonance backscatter area), the tilt MTF can be calculated from simple analytical expressions (Wright 1968):

$$T_{\text{tilt},\parallel}(\kappa) = \begin{cases} \dfrac{4i\kappa_x \cot\theta_0}{1 + \sin^2\theta_0}, & \text{VV polarization} \\[3mm] \dfrac{8i\kappa_x}{\sin^2\theta_0}, & \text{HH polarization} \end{cases} \tag{3.45}$$

Equations (3.45) are true for large sea water dielectric constants (which are of the order of 80 for electromagnetic wavelength exceeding approximately 10 cm (see Figures 3.1 and 3.2) and the ripple spectrum $W_\xi(\kappa) \propto \kappa^{-4}$.

Consider a simple situation with a single wave propagating in the range direction. The corresponding "frozen" spatial wave field is

$$\zeta(x) = A_0 \cos(\kappa_0 x + \varphi) = \frac{1}{2}A_0 \, e^{i(\kappa_0 x + \varphi)} + \text{c.c} \tag{3.46}$$

where φ is the arbitrary phase. Having substituted into Eqn. (3.39),

$$A_\zeta(\kappa) = \frac{1}{2}A_0\delta(\kappa - \kappa_0) \tag{3.47}$$

and integrating over κ, we get

$$\frac{\delta\sigma_0}{\langle\sigma_0\rangle} = 1 + \frac{1}{2}\left[T(\kappa_0)A_0\,\mathrm{e}^{\mathrm{i}(\kappa_0 x+\varphi)} + \text{c.c.}\right] = 1 + \frac{1}{2}\left[\frac{T(\kappa_0)}{\mathrm{i}\kappa_0}\mathrm{i}\kappa_0 A_0\,\mathrm{e}^{\mathrm{i}(\kappa_0 x+\varphi)} + \text{c.c.}\right] \tag{3.48}$$

Taking into account that $T(\kappa_0)/ik_0$ is a real value, we write

$$\frac{\delta\sigma_0}{\langle\sigma_0\rangle} = 1 + \frac{1}{2}\frac{T(\kappa_0)}{\mathrm{i}\kappa_0}\left(\mathrm{i}\kappa_0 A_0\mathrm{e}^{\mathrm{i}(\kappa_0 x+\varphi)} + \text{c.c}\right) = 1 - \frac{T(\kappa_0)}{\mathrm{i}\kappa_0}\kappa_0 A_0 \sin(\kappa_0 x + \varphi) \tag{3.49}$$

Let a surface wave be ripples with the wavelength $\Lambda_0 = 2\pi/\kappa_0 = 200$ m and amplitude 3 m. For the incidence angle $\theta_0 = 23°$ of the vertically polarized electromagnetic field (Figure 3.9a) we find $T/\mathrm{i}\kappa_0 \approx 10$, and from it we draw a conclusion that

$$\frac{\delta\sigma_{0,\text{tilt}}}{\langle\sigma_0\rangle} \approx 1 - \sin(\kappa_0 x + \varphi) \tag{3.50}$$

In the case of an azimuthally directed wave, the tilt modulation is highly weak as it can be easily concluded from the small values of $T_{\text{tilt}\parallel}/\mathrm{i}\kappa$ for the whole interval of incidence angles.

Speaking about the MTF hydrodynamic part, as mentioned previously, the hydrodynamic theory is still under development. According to the existing theoretical estimate (Alpers and Hasselmann 1978)

$$T_{\text{hydr}} = -4.5\kappa\Omega\frac{\Omega-\mathrm{i}\mu}{\Omega^2+\mathrm{i}\mu^2}\cos^2\varphi_0 \tag{3.51}$$

Here φ_0 is the angle between the horizontal projection of radar look direction and the surface wave direction, and μ^{-1} is the so-called "relaxation time constant" defined by the phase shift between the maximum of the short-wave spectral energy and the large wave crest. Numerous investigations, visual ones included, indicate that the phase shift has value between 0 and $\pi/2$. To simplify the expression we introduce $\mu = 0$ (that corresponds to zero phase shift) and obtain

$$T_{\text{hydr}} \approx -4.5\kappa\cos^2\varphi_0 \tag{3.52}$$

For the same swell wave with amplitude 3 m and wavelength 200 m in the range direction, it follows from Eqn. (3.52) that

$$\frac{\delta\sigma_{0,\text{hydr}}}{\langle\sigma_0\rangle} \approx 1 + 0.4\cos(\kappa_0 x + \varphi) \tag{3.53}$$

Thus, according to the theoretical estimate, the tilt modulation (more than twice) exceeds the hydrodynamic one considerably.

MTF has been investigated in a series of field experiments (see, e.g. Plant et al. 1983, Schmidt et al. 1993), with modules and MTF phase figures essentially varying for one and the same experiment. Alongside the MTF parameters, the correlation degree between σ_0 and ς was investigated, the latter being was found considerably low, even though in the accordance with the two-scale model scattering theory their relation is close to linear (Plant et al. 1983). The MTF parameters, defined in the course of the subsequent experiments (Schmidt et al. 1995, Hauser and Caudal 1996), also show very large discrepancy; besides, both works point out significant discrepancies between hydrodynamic modulation theory and experiment. In particular, Hauser and Caudal (1996) find that the modulus of the hydrodynamic transfer function is several times larger (by a factor 2–12) than the theoretical value proposed in previous works and 1.5–2.5 larger than the experimental values reported in recent papers.

Most probably, these discrepancies are caused not only by the underdeveloped theory, but also by ocean surface inhomogeneities with wide range of scales, whose presence is testified by the airborne and space-borne radar imaging (see Chapter 1). Various oceanic processes, such as internal waves, local currents and surface-active films, modify surface roughness, first, its small-scale part. Besides, centimetre radar signal scattered by sea surface almost always includes more or less intense "wind noise". This noise is the result of wind speed fluctuations, which are manifested on the surface as centimetre ripple spots of various contrast and size, and these ripples further find reflection in the radar signal (Zhydko et al. 1983). Figure 3.10 shows the normalized cross-correlation function (correlation coefficient) of microwave radar signal intensity and wind speed; the data were collected in the course of a sea experiment with ship-borne radar (Zhydko and Ivanova 2001). It is clear from Figure 3.10 that the maximum correlation coefficient between radar signal and wind speed is approximately 0.9 and decreases two times at temporal lag of about 1 min.

Therefore, apart from the large-scale waves, σ_0 is modulated by these foreign processes. Evidently, it breaks the linear relation between σ_0 and ζ, and generates certain peculiarities in the surface image spectrum, which are not described by MTF.

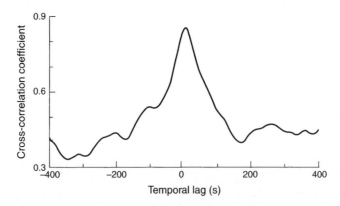

Figure 3.10 Cross-correlation coefficient of wind speed and 3.2 cm wavelength radar signal [Zhydko and Ivanova 2001].

All this means that MTF can theoretically be used to retrieve the roughness spectrum from the radar image as well as to obtain the data on the interaction between long and short gravity–capillary waves. However, respective experiments should be carried out on the sufficiently homogeneous roughness, i.e. only slightly disturbed by foreign processes. As shown in Figure 2.6, the condition is especially difficult to meet exactly in the coastal area, where field experiments to estimate MTF parameters are mostly carried out.

Besides the radar cross section, the large-scale roughness modulates the backscattered microwave signal frequency as well owing to the orbital movement of backscattering ripples on the long waves. Hence microwave Doppler spectra differ sufficiently from the HF ones. Though microwave Doppler spectrum is similarly based on two lines, the lines far from always can be resolved through ripples orbital movements. Doppler spectral width in case of stationary radar is determined by width of the small-scale orbital velocity spectrum. The backscattered microwave signal Doppler spectrum is considered in more detail in the next chapter.

Everything said holds within applicability of the sea surface two-scale model. When does it apply then? First of all, incidence angle should not be too small (i.e. the incidence should not be too steep). Otherwise, the resonance wave number gets into the large-scale part of the roughness spectrum, where the perturbation theory does not work. The respective border incidence angle depends on the surface state; in moderate winds it is 10–20°. Furthermore, it is known that at small depression angles (about 15° or less) the resonant backscattering turns out to be significantly lower than the one caused by breaking wave elements – this mainly applies to horizontally polarized field (see Section 3.3). We can clearly see, particularly in Figure 3.8b, that the experimental values σ_0 at small grazing angles exceed those computed based on the two-mode pattern.

Thus, in winds of up to approximately $20\,\text{m s}^{-1}$, the two-scale model is applicable within a wide incidence angle range of 20–75°. The two-scale model was proposed in its original form in Kuryanov (1962) for the acoustic radiation scatter, and then appeared evolutionized, in its more general form in Fuks (1966), Bass et al. (1968), Wright (1968) and Bass and Fuks (1979). It proved to be quite fruitful. It has been used to explain numerous experimental facts and recover abundant theoretical data, later supported experimentally.

3.3 BACKSCATTERING AT SMALL INCIDENCE ANGLES

Radio oceanography also deals with the small incidence angles. The typical radar tool working in this angle range is radar altimeter, looking in nadir. As its name suggests, altimeter is commonly used to measure the altitude of a radar platform above the surface. With space altimeters the data on Earth's geoid form and oceanic topography are obtained. Apart from these, altimeter has other features to offer; it also determines reflected pulse shape and the amount of scattered power. The reflected pulse shape allows to make an estimate of the significant wave height H_s with the accuracy of 10–20 cm, which is important by itself, besides, the data on H_s and NRCS σ_0 are both necessary for the retrieval of near-surface wind speed.

Besides conventional altimeters, there are scanning radar altimeters used in steep incidence regime providing panoramic imaging of the ocean surface (Walsh et al. 1985, Walsh and Vandemark 1998, Wright et al. 2001).

3.3.1 Radar cross section at small incidence angles

As it has been mentioned above, at moderate incidence angles large gravity waves manifest themselves in radar imagery of the sea surface indirectly, i.e. by spatio-temporal modifications of the resonance gravity–capillary ripples, induced by these large waves. Moving into the steep incidence area (almost vertical incidence angles), utmost importance is given to the direct reflection from the large sea waves, as in this case there is a chance of specular (or quasi-specular, i.e. close to specular) reflection from large waves. In this case the small perturbation method becomes unfeasible, because the wavelength that meets the Bragg condition, falls within the large-scale part of sea roughness spectrum, and for large surface waves field $\zeta(\vec{r}, t)$ and microwaves, the expression is always $k\sigma_\zeta \gg 1$.

For the small incidence angles $\theta_0 = \pi/2 - \psi_0$, the reflected field characteristics are calculated based on the Kirchoff method, where the source is the Green equation:

$$u(\vec{r}_0) = \frac{1}{4\pi} \oint \left[u(\vec{r}_S)\frac{\partial}{\partial n}\left(\frac{\mathrm{e}^{\mathrm{i}kR}}{R} \right) - \frac{\mathrm{e}^{\mathrm{i}kR}}{R}\frac{\partial u}{\partial n}(\vec{r}_S) \right] \mathrm{d}\vec{r}_S \qquad (3.54)$$

Here $u(\vec{r}_0)$ is the spectral amplitude of the scattered field (scalar, for simplicity) in the observation point, and the integration is over a closed surface S, which specifies the field values and its derivative over the normal to S; the value $R = \left| \vec{r}_0 - \vec{r}_S \right|$ is the distance between the observation point and the current point on the surface.

It is to be emphasized that we speak only about the scattered field, as Eqn. (3.54) does not include the term corresponding to the field coming directly from the source.

Let S be a hemisphere resting on a surface with roughness elements described by the function $\zeta(\vec{r}, t)$ (Figure 3.11); moreover, there exists a small electromagnetic absorption

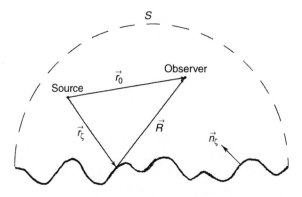

Figure 3.11 Geometrical configuration of the electromagnetic scattering problem.

inside S. We will set the hemisphere radius as infinite and consider the infinity condition according to which the field in infinity is a system of outgoing waves. As a result the integral over the hemisphere becomes zero and it is solely $\zeta(\vec{r},t)$ that contributes to the scattered field.

Therefore, the scattered field in the given space point is defined through the value of the field itself and its normal derivative on the surface $\zeta(\vec{r},t)$ should be set with due consistency. The way to set these values makes well the essence of the Kirchoff method.

The main idea of the method is the assumption that at each points of a rough surface, the incident and scattered fields (respectively, u_0 and u) are connected by relations applicable for the incidence of a *plane* wave onto the *plane* border of two media divisions:

$$u\left(\vec{r}_\zeta\right) = V\left(\vec{r}_\zeta\right)u_0\left(\vec{r}_\zeta\right) \tag{3.55a}$$

$$\frac{\partial u}{\partial n}\left(\vec{r}_\zeta\right) = -V\left(\vec{r}_\zeta\right)\frac{\partial u_0}{\partial n}\left(\vec{r}_\zeta\right) \tag{3.55b}$$

where V is the reflection coefficient (for electromagnetic field it is the Fresnel coefficient dependent on the incidence wave polarization). Relations (3.55) mean that the surface is considered as locally flat, that is to say when evaluating the reflection it can be replaced by tangent plane at each point. It evidently brings about certain restrictions for the reflecting surface characteristics and electromagnetic wave incidence (grazing) angles. What are the restrictions? First of all, the roughness elements are to be characterized by small degree of curvature, or, in other words, by a fairly big radius of curvature (Brekhovskikh 1952):

$$k\,R\sin^3\psi \gg 1 \tag{3.56}$$

where R is the average curvature radius and ψ is the local grazing angle. Besides, an important condition is the absence of shading of surface elements by the other ones and multiplicative scattering, when the field has been reflected from one inhomogeneous element to another before it reaches the observation tool.

Substitution of Eqns (3.55a) and (3.55b) into Eqn. (3.54) yields the following expression for the scattered field:

$$u = \frac{1}{4\pi}\int V\left(\vec{r}_\zeta\right)\frac{\partial}{\partial n}\left\{\frac{e^{-ikR}}{R}u_0\left(\vec{r}_\zeta\right)\right\}\,d\vec{r}_\zeta \tag{3.57}$$

After a sequence of formal transformations (see for more details, Bass and Fuks (1979) and Rytov et al. (1989)) carried out at

$$R \gg \lambda, \quad k^2\sigma_\zeta^2 \gg 1 \tag{3.58}$$

for the backscattered field of the spherical wave $(A_0/R)\exp(ikR)$, incident on the rough, on average, flat area S_0, expression (3.57) gives

$$u = -\frac{iV_\perp A_0}{\lambda R_0^2 \cos\theta_0} \int_{S_0} \exp\left\{2ik\left[R(\vec{r}) - \zeta(\vec{r})\cos\theta_0\right]\right\} d\vec{r} \qquad (3.59)$$

Here R_0 and θ_0 are the range and incidence angle on the plane of the central ray, $R(\vec{r})$ is the distance from the observation point to current point on the average plane XY over which the integration is performed within S_0 and $V_\perp$ is the reflection coefficient for the vertical incidence of a flat wave onto a flat border. The presence of reflection coefficient for vertical incidence in Eqn. (3.59) is accounted for by the dominating contribution into backscatter by large wave elements perpendicular to the incident ray. As is easily seen, at $k\sigma_\zeta^2 \ll R$, the $R(\vec{r}) - \zeta(\vec{r})\cos\theta_0$ value in the exponent of the integrand of Eqn. (3.59) is nothing but the distance between the antenna to the current point on the surface $\zeta(\vec{r})$.

On the basis of Eqn. (3.59) we find the NRCS

$$\sigma_0 = \frac{4\pi R_0^2}{S_0}\lim_{R_0 \to \infty}\left(\frac{\langle u \rangle^2}{|u_0|^2}\right) \qquad (3.60)$$

(as been determined in Chapter 1, see Eqn. (1.25)) for the quasi-specular scatter area. Substituting $u_0 = (A_0/R_0)\exp(ikR_0)$ and expression (3.59) for the backscattered field u into Eqn. (3.60), we have

$$\sigma_0 = \frac{4\pi|V_\perp|^2}{S_0\lambda^2\cos^2\theta_0}\lim_{R_0 \to \infty}\langle K \rangle \qquad (3.61a)$$

$$\langle K \rangle = \iint_{S_0} d\vec{r}'d\vec{r}'' \exp\left\{2ik\left[R(\vec{r}') - R(\vec{r}'')\right]\right\}\left\langle \exp\left\{-2ik\cos\theta_0\left[\zeta(\vec{r}') - \zeta(\vec{r}'')\right]\right\}\right\rangle$$

$$(3.61b)$$

It is easy to notice that part of the integrand in Eqn. (3.61b), the expression in the angle brackets, to be exact, is nothing but a characteristic function of two variables Θ_2, for which at normal distribution of random field $\zeta(\vec{r})$ the following formula holds true:

$$\Theta_2 = \exp\left\{-4k^2\cos^2\theta_0\left[B_\zeta(0) - B_\zeta(\vec{\rho})\right]\right\} \qquad (3.62)$$

where $B_\zeta(\vec{\rho}) = B_\zeta(\vec{r}' - \vec{r}'')$ is the correlation function of the field ζ with $\vec{\rho} = \{\rho_x, \rho_y\}$.

The condition $k^2\sigma_\zeta^2 \gg 1$ can always stand as realized for microwaves and large sea waves. Thereforethe function Θ_2 has a sharp peak at $\rho = 0$, and the main contribution to the integral (3.61b) is made by small range of point $\rho = 0$, within which

$$\Theta_2 \approx \exp\left\{-4k^2\cos^2\theta_0\left[-\frac{1}{2}\frac{\partial^2 B_\zeta}{\partial\rho_x^2}\bigg|_{\vec{\rho}=0}\rho_x^2 - \frac{\partial^2 B_\zeta}{\partial\rho_x\partial\rho_y}\bigg|_{\vec{\rho}=0}\rho_x\rho_y - \frac{1}{2}\frac{\partial^2 B_\zeta}{\partial\rho_y^2}\bigg|_{\vec{\rho}=0}\rho_y^2\right]\right\}$$

(3.63)

The Wiener–Khinchin theorem (see Chapter 1) yields

$$-\frac{\partial^2 B_\zeta}{\partial\rho_x^2}\bigg|_{\vec{\rho}=0} = \int d\vec{\kappa}\,\kappa_x^2 W_\zeta(\vec{\kappa}) = \sigma_x^2$$

(3.64a)

$$-\frac{\partial^2 B_\zeta}{\partial\rho_y^2}\bigg|_{\vec{\rho}=0} = \int d\vec{\kappa}\,\kappa_y^2 W_\zeta(\vec{\kappa}) = \sigma_y^2$$

(3.64b)

$$-\frac{\partial^2 B_\zeta}{\partial\rho_x\partial\rho_y}\bigg|_{\vec{\rho}=0} = \int d\vec{\kappa}\,\kappa_x\kappa_y W_\zeta(\vec{\kappa}) = \sigma_{xy}$$

(3.64c)

Here $W_\zeta(\vec{\kappa})$ is the spectrum of surface elevations $\zeta(\vec{r})$, and σ_x^2 and σ_y^2 are the variance of surface slopes along x and y, respectively. As for σ_{xy}, as easily displayed, $\sigma_{xy} = 0$ when the general direction of wave propagation coincides with one of the coordinate axes. Note that expressions (3.64) include the spectrum $W_\zeta(\vec{\kappa})$, and not $\hat{W}_\zeta(\vec{\kappa})$, and integration is over wave vectors $\vec{\kappa}$, corresponding to the true directions of wave propagation.

We consider the central ray incidence plane as parallel to the coordinate plane *XZ* (see Figure 1.1), and the general direction of roughness propagation as coinciding with the axis x or opposite to it.

Notably, at $R \to \infty$ it is valid that

$$R = R_0 + (x - x_0)\sin\theta_0$$

(3.65)

and, consequently, the integrand in Eqn. (3.61b) depends only on the difference variables ρ_x, ρ_y. If the condition $\Delta x, \Delta y \gg (\kappa\sigma_x, k\sigma_y)^{-1}$ is met, where $\Delta x, \Delta y$ is the linear dimensions of the cell and $\sigma_x, \sigma_y \approx 0.1 - 0.2$, it is possible to integrate over variables ρ_x, ρ_y infinitely; therefore,

$$\langle K\rangle = S_0\int_{-\infty}^{\infty}d\rho_x\exp\left(-2k^2\sigma_x^2\cos^2\theta_0\rho_x^2 + 2ik\sin\theta_0\rho_x\right)$$
$$\times \int_{-\infty}^{\infty}d\rho_y\exp\left(-2k^2\sigma_y^2\cos^2\theta_0\rho_y^2\right)$$

(3.66)

The integrals in Eqn. (3.66) are easily obtained. We then obtain the following formula for the NRCS:

$$\sigma_0 = \frac{|F_0|^2}{2\sigma_x\sigma_y}\sec^4\theta_0 \exp\left(-\frac{\tan^2\theta_0}{2\sigma_x^2}\right) \tag{3.67}$$

(we have replaced the reflection coefficient $V_\perp$ by the Fresnel one). As easily noticed, σ_0 has turned out proportional to the probability degree of such a surface slope in the central ray incidence spot that provides the specular reflection of the ray.

Note that formula (3.67) was established only on the basis of large-scale roughness, and we did not take into consideration the small-scale ripples causing diffuse scattering that reduces the scatter cross section. The ripples influence is estimated with the help of effective reflection coefficient F_0^{eff}, introduced instead of Fresnel coefficient (Barrick 1974, Miller et al. 1984, Miller and Veigh 1986). Figure 3.12 (from Valenzuela 1978) shows the dependence of effective reflection coefficient on the wind speed.

For isotropic roughness $\sigma_x^2 = \sigma_y^2$,

$$\sigma_0^{\text{isotr}} = \frac{|F_0^{\text{eff}}|^2}{\sigma_{\text{sl}}^2}\sec^4\theta_0 \exp\left(-\frac{\tan^2\theta_0}{\sigma_{\text{sl}}^2}\right) \tag{3.68}$$

where $\sigma_{\text{sl}}^2 = \sigma_x^2 + \sigma_y^2$ is the total variance of slopes. Figure 3.13 (from Valenzuela 1978) gives theoretical and experimental values of NRCS at incidence angles $0 \le \theta_0 \le 20^0$. Theoretical curves for isotropic roughness have been obtained with regard to the dependence $|F_0^{\text{eff}}|^2$ on the wind speed as displayed in Figure 3.12.

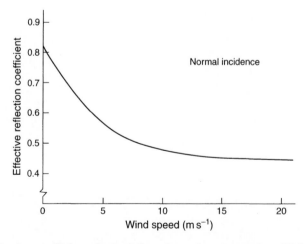

Figure 3.12 Reflection coefficient obtained from 3 cm electromagnetic radiation measurements [Valenzuela 1978, originally cited from Barrick 1974].

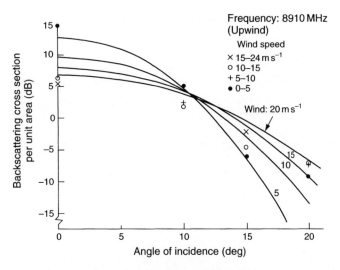

Figure 3.13 Comparison of measured and theoretical (with the results given in Figure 3.12 taken into account) cross sections per unit area for various winds (from Valenzuela [1978]).

When looking at nadir ($\theta_0 = 0$) with provision for small-scale roughness, expression (3.68) takes on a very simple form:

$$\sigma_0 = \frac{\left|F_0^{\mathrm{eff}}\right|^2}{2\sigma_x\sigma_y} \qquad (3.69)$$

The example of near-nadir radar probing with the airborne scanning radar altimeter is given in Figure 3.14 (quoted in Wright et al. 2001). The radar simultaneously measured the backscattered power at its 36 GHz ($\lambda = 8.3$ mm) operational frequency and the range

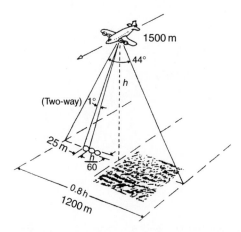

Figure 3.14 Measurement geometry of the scanning radar altimeter [Wright et al. 2001].

to the sea surface. In this work the sea surface directional wave spectrum was measured for the first time in all quadrants of a hurricane inner core over open water.

The connection between the radar cross section and the ocean surface state is employed in the radar altimetric measurement of the wind speed over the ocean. Figure 3.15 (from Karaev et al. 2006a) compares the values σ_0 and the wind speed values acquired independently. Multitude of points (overall number is 2860) comprises σ_0 data, acquired with the radar altimeter of the European satellite ERS-2 in 1995–2000, and the hydrographic pitch-and-roll buoy data on the wind speed. The data on the radar cross section, on the one hand, and wind speed, on the other, are spread apart in space and in time not more than 50 km and 30 min, respectively.

As suggested by Eqns (3.68) and (3.69), the radar cross section, in general, decreases with increase in wind speed. However considerable spread of points in Figure 3.15 indicates that there is no single-valued correspondence between σ_0 and the near-surface wind speed, which is quite explicable from the physical standpoint. The matter is that the radar cross section expression includes large-scale roughness slopes and, according to Chapter 1, the slopes correspond one to one to the wind speed only at fully developed windsea with no swell. In case of swell and (or) undeveloped windsea on the surface at an unknown wave fetch value, the large-scale wave slopes may differ considerably at the same near-surface wind speed, and, consequently, the values σ_0 will also be different. For this reason the accuracy of one-parameter (i.e. built only with evaluation of σ_0) algorithms of near-surface wind speed measurement is rather low; as a rule it does not exceed $2\,\mathrm{m\,s^{-1}}$.

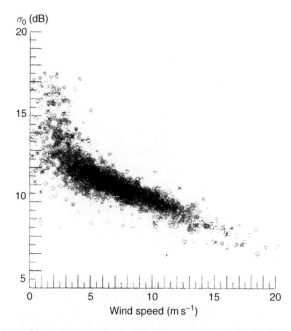

Figure 3.15 The dependence of normalized radar cross section on wind speed. The data collected from ERS-2 radar altimeter in 1996–2000 (in total 2860 points are given) [Karaev et al. 2006a].

The accuracy reaches $1.4 - 1.5$ m s^{-1} with the use of two-parameter algorithms; when together with σ_0 the measurements include the large-scale roughness height statistically connected with slopes and at the same time take into account the regional specifics of water areas under study (Karaev et al. 2002a, 2002b). (The large-scale roughness height $H_s = 4\sigma_z$ where σ_z is the RMS of surface elevations, is measured by an altimeter independently of σ_0, according to the leading-edge time of the backscattered pulse (Stewart 1985).)

However cardinal enhancement in ocean surface wind speed retrieval at near-nadir probing from the space can only be obtained with the possibility of directly measuring the surface slopes in the broad swath (Karaev et al. 2005, 2006b). At the same time the significant advance in the wind speed-retrieval problem solution has been outlined with the experiments performed with space SAR (see Section 6.8.3).

3.3.2 Doppler spectrum at small incidence angles

In this section we will deduce the formulae linking the parameters of Doppler spectrum for the radar signal backscattered from the sea surface at small incidence angles and roughness characteristics, as well as appraise the potential of this probing method. We will proceed with Eqn. (3.59), and first obtain on its basis the backscattered field correlation function

$$B_u(\tau) = \langle u(t)u^*(t+\tau)\rangle \tag{3.70}$$

and then switching over to the Doppler spectrum with the help of Fourier transformation, we will find its parameters – the width and position of the peak.

In this case as the correlation function is of the main interest for us, we will omit the coefficients before the integral in Eqn. (3.59) and write

$$B_u(\tau) \propto \iint d\vec{r}'d\vec{r}'' \, \phi\left(\vec{r}' - \vec{r}_0\right)\phi\left(\vec{r}'' - \left(\vec{r}_0 + \vec{\rho}_0\right)\right)$$

$$\times \exp\left\{2ik\left[R\left(\vec{r}', \vec{r}_0\right) - R\left(\vec{r}'', \vec{r}_0 + \vec{\rho}_0\right)\right]\right\}$$

$$\times \left\langle \exp\left\{-2ik\cos\theta_0\left[\varsigma\left(\vec{r}', t\right) - \varsigma\left(\vec{r}'', t+\tau\right)\right]\right\}\right\rangle \tag{3.71}$$

where $\vec{r}_0 = \{x_0, y_0\}$, $\vec{\rho}_0 = \{0, V\tau\}$ and the function

$$\phi = \exp\left\{-\left[\left(\frac{x - x_0}{\Delta x/2}\right)^2 - \left(\frac{y - y_0}{\Delta y/2}\right)^2\right]\right\} \tag{3.72}$$

gives the amplitude value of the field reflected to the antenna by the point $\zeta(x, y)$ on the rough surface depending on the distance from the point (x, y) to the centre (x_0, y_0) of the illuminated spot measured on the surface XY. Δx, Δy values are characteristic linear size of an illuminated spot taking into account the twofold operation of the antenna pattern at radar signal transmission and receiving.

Furthermore, we will consider as accomplished the condition

$$\frac{k(\Delta x)^2}{R_0} \ll 1 \tag{3.73}$$

which in fact means that we neglect the influence of the rays divergence in the vertical plane on the field phase characteristics. As for the azimuth plane, we cannot make the same allowance, as it is evident in advance that at a fairly high speed of radar carrier it is exactly the angular width of the azimuth antenna pattern that determines the reflected signal Doppler spectrum width. Thus,

$$R\left(\vec{r}, \vec{r}_0\right) = R_0 + (x - x_0)\sin\theta_0 + \frac{1}{R_0}(y - y_0)^2 \tag{3.74}$$

As before with the similar integral (3.61b), the integrand of Eqn. (3.71) also has the characteristic function Θ_2; however, the values of the random variable $\zeta(\vec{r}, t)$ are not only spaced apart but also lag in time. Hence the expression for Θ_2 differs from Eqn. (3.63) by the presence of the second derivative over τ in the exponent, and mixed derivatives over ρ_x, τ and ρ_y, τ:

$$\Theta_2 = \exp\left[-4k^2\cos^2\theta_0\left(|K_{xx}|\rho_x^2 + |K_{yy}|\rho_y^2 + |K_{tt}|\tau^2 - K_{xy}\rho_x\rho_y - K_{xt}\rho_x\tau - K_{yt}\rho_y\tau\right)\right] \tag{3.75}$$

$$K_{\xi\xi} = 0.5\frac{\partial^2 B_\zeta}{\partial\xi^2}, \quad K_{\xi\eta} = \frac{\partial^2 B_\zeta}{\partial\xi\,\partial\eta}, \quad \xi, \eta = \rho_x, \rho_y, \tau \quad (\xi \neq \eta)$$

All derivatives are taken at $\rho_x, \rho_y, \tau = 0$.

Now that the simplifications and assumptions for the integrand in Eqn. (3.71) are fully determined, we can find the correlation function $B_u(\tau)$, and then the signal Doppler spectrum. After a number of simple, but rather tedious calculations, we get a Gaussian-shaped Doppler spectrum with the peak at frequency f_{sh} and width Δf at the level -10 dB (Kanevsky and Karaev 1996a):

$$f_{sh} = -\frac{\sin\theta_0}{\lambda}\left[\beta_{xt} + \beta_{xy}\frac{\gamma_{yt}\cos^2\theta_0 - 0.09V\delta_{y(2)}^2}{\alpha_{xy}\cos^2\theta_0 + 0.09\delta_{y(2)}^2}\right] \tag{3.76}$$

$$\Delta f = \frac{4\sqrt{\ln 10}}{\lambda}\left[0.09V^2\delta_{y(2)}^2 + \alpha_{xt}\cos^2\theta_0 - \frac{\left(\gamma_{yt}\cos^2\theta_0 - 0.09V\delta_{y(2)}^2\right)^2}{\alpha_{xy}\cos^2\theta_0 + 0.09\delta_{y(2)}^2}\right]^{1/2} \tag{3.77}$$

$$\alpha_{xt} = 4|K_{tt}| - \frac{K_{xt}^2}{|K_{xx}|}, \quad \beta_{xt} = \frac{K_{xt}}{|K_{xx}|}, \quad \alpha_{xy} = 4|K_{yy}| - \frac{K_{xy}^2}{|K_{xx}|},$$

$$\beta_{xy} = \frac{K_{xy}}{|K_{xx}|}, \quad \gamma_{yt} = 2K_{yt} + \frac{K_{xy}K_{xt}}{|K_{xx}|}$$

where $\delta_{y(2)}$ is the width of two-way antenna pattern in the azimuth plane at the level 0.5. The derivatives from $B_\zeta(\rho_x, \rho_y, \tau)$ can be found through spatial and temporal spectra of elevations:

$$-\left.\frac{\partial^2 B_\zeta}{\partial\rho_x^2}\right|_{\vec{\rho},\tau=0} = \int d\vec{\kappa}\,\kappa_x^2 W_\zeta(\vec{\kappa}) \tag{3.78a}$$

$$-\left.\frac{\partial^2 B_\zeta}{\partial\rho_y^2}\right|_{\vec{\rho},\tau=0} = \int d\vec{\kappa}\,\kappa_y^2 W_\zeta(\vec{\kappa}) \tag{3.78b}$$

$$-\left.\frac{\partial^2 B_\zeta}{\partial\rho_x\partial\rho_y}\right|_{\vec{\rho},\tau=0} = \int d\vec{\kappa}\,\kappa_x\kappa_y W_\zeta(\vec{\kappa}) \tag{3.78c}$$

$$-\left.\frac{\partial^2 B_\zeta}{\partial\tau^2}\right|_{\vec{\rho},\tau=0} = \int_0^\infty d\omega\,\omega^2 G_\zeta(\omega) \tag{3.78d}$$

Remember that the right-hand side of the first two equations are variance of the surface slopes along the x- and y-axis, respectively, and the right-hand side of the fourth equation is the surface vertical shift speed variance. The mixed derivatives over the spatial and temporal coordinates are as follows:

$$\left.\frac{\partial^2 B_\zeta}{\partial\rho_x\partial\tau}\right|_{\vec{\rho},\tau=0} = \int d\vec{\kappa}\int_0^\infty d\omega\,\kappa_x\omega\Psi_\zeta(\vec{\kappa},\omega) \tag{3.79a}$$

$$\left.\frac{\partial^2 B_\zeta}{\partial\rho_y\partial\tau}\right|_{\vec{\rho},\tau=0} = \int d\vec{\kappa}\int_0^\infty d\omega\,\kappa_y\omega\Psi_\zeta(\vec{\kappa},\omega) \tag{3.79b}$$

where $\Psi_\zeta(\vec{\kappa},\omega) = W_\zeta(\vec{\kappa})\delta[\omega - \Omega(\kappa)]$ is the spatio-temporal spectrum of the surface eleva-tions $\zeta(\vec{r},t)$, and the temporal frequency Ω and the wave number κ are interconnected by the dispersion relation for gravity waves: $\Omega^2 = \kappa g$ Integrating the right-hand sides of Eqns (3.79a) and (3.79b) over ω, we obtain

$$\left.\frac{\partial^2 B_\zeta}{\partial \rho_x \partial \tau}\right|_{\vec{\rho},\tau=0} = g^{1/2} \int d\vec{\kappa}\, \kappa_x \kappa^{1/2} W_\zeta(\vec{\kappa}) \tag{3.80a}$$

$$\left.\frac{\partial^2 B_\zeta}{\partial \rho_y \partial \tau}\right|_{\vec{\rho},\tau=0} = g^{1/2} \int d\vec{\kappa}\, \kappa_y \kappa^{1/2} W_\zeta(\vec{\kappa}) \tag{3.80b}$$

We can integrate the right-hand sides of Eqns. (3.78) and (3.80) after we turn to polar coordinates and use the gravity wave spatial spectral adopted above (see Chapter 1).

Thus, formulae (3.76) and (3.77) connect the roughness spectrum and radar signal Doppler spectrum parameters in the quasi-specular area.

Before we place here the results obtained with these expressions, let us review the following simple concepts. In the area of small incidence angles, the backscattered field is made of plane elements of the surface perpendicular (or nearly perpendicular) to the incident ray. In the hypothetical case of monochromatic roughness these planes, formed by certain segments of wave profile, move horizontally (parallel to themselves) together with the wave characterized by phase velocity v_{ph}. The reflected signal will be then monochromatic with the frequency shifted by the value $f_{sh} = -(2v_{ph}/\lambda)\sin\theta_0\,\cos\varphi_0$, where φ_0 is the angle between the projection of the probing ray onto the horizontal plane and the wave propagation direction. As the roughness spectrum broadens, the system of surface specular elements produced by the sum of waves with various phase velocities and occasional mutual phase shifts, becomes random, and the lifetime of each segment becomes finite; moreover, the shorter the lifetime, the broader the spectrum. Besides, the segments acquire random vertical velocities. All this broadens the reflected signal spectrum, and Doppler shift of the spectrum peak f_{sh} is defined by the average phase velocity of the sea waves. This phase velocity can be expressed through the values from expressions (3.76) and (3.77), by assuming $\varphi_0 = 0°$ or $\varphi_0 = 180°$. In this case $\beta_{xy} = 0$, and at $\varphi_0 = 180°$ (upwind look) we obtain the following formula for the Doppler frequency shift:

$$f_{sh} = \frac{\sin\theta_0}{\lambda}\frac{K_{xt}}{|K_{xx}|} = \frac{2\,\sin\theta_0}{\lambda}\frac{K_{xt}}{\sigma_x^2} \tag{3.81}$$

and consequently

$$\langle v_{ph}\rangle = \frac{K_{xt}}{\sigma_x^2} \tag{3.82}$$

Thus, the average phase velocity for the roughness generally propagating along or against the X-axis equals

$$\langle v_{ph} \rangle = \frac{g^{1/2} \int d\vec{\kappa} \; \kappa_x \kappa^{1/2} W_\zeta(\vec{\kappa})}{\int d\vec{\kappa} \; \kappa_x^2 W_\zeta(\vec{\kappa})} \tag{3.83}$$

This formula can prove useful for theoretical studies.

For different roughness types the spectra breadth, on the one hand, and the wave phase velocities, on the other, correlate in different ways. This means that the situations matching different roughness types can be basically grouped apart in the "signal Doppler spectrum width vs shift" coordinate system.

Figure 3.16a and b shows the calculation results for the electromagnetic wave $\lambda = 3$ cm and incidence angle $\theta_0 = 10°$; the probing is performed at $\varphi_0 = 180°$, i.e. in the upwind radar look direction. Figure 3.16a features a stationary platform (radar carrier speed is $V = 0$) at the antenna pattern width performing at half-capacity $\delta_y = 3°$, and Figure 3.16b refers to another case: $V = 200$ m s^{-1} (the aircraft carrier), $\delta_y = 0.5°$ (along-fuselage antenna).

The bold curve stands for the fully developed roughness, and the points on it mark the wind speed values. To the points on the bold curve there approach thin curves, each of them shows the developing roughness at the given wind speed. The thin curves start with dimensionless fetch value $\tilde{X}_w \approx 2800$ and reach the bold curve at $\tilde{X}_w = 20,000$. Recall (see Chapter 1) that at $\tilde{X}_w = 20,000$, the developing roughness becomes fully developed.

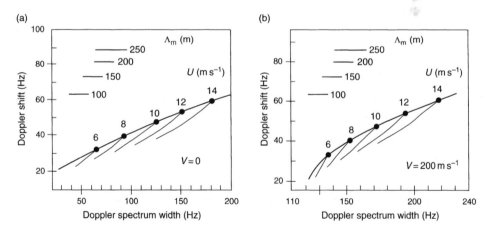

Figure 3.16 Relation between Doppler spectrum width and its maximum shift: (a) a stationary platform and (b) an aircraft ($V = 200$ m s^{-1}). The curve in bold stands for the fully developed roughness, the points on it mark the wind speed values. To the points on the bold curve there approach thin curves, each of them shows the developing roughness at the given wind speed. Horizontal line segments in the top part of the figures correspond to swell, its parameter is the dominant wavelength Λ_m (for further details, see text).

The horizontal line segments in the upper part of Figure 3.16a and b correspond to swell, and its parameter is the dominant wavelength Λ_m. Calculations were based on the spectrum (2.28), where the swell height values $H = 4\sigma_\zeta$ were defined within the following limits: $0.88 - 1.34$ m ($\Lambda_m = 100$ m); $1.44 - 2.2$ m ($\Lambda_m = 150$ m); $2.0 - 2.69$ m ($\Lambda_m = 200$ m); and $2.24 - 3.37$ m ($\Lambda_m = 250$ m). As we see, for the given value of the swell-dominant wavelength, the Doppler shift is almost independent of its height. However, the larger the swell, the higher the variance of the surface vertical shifts and, consequently, the broader the reflected signal spectrum.

The information above describes the case where there is only one wave type on the sea surface (one-mode roughness). If apart from windsea waves there is also one or more swell systems, the picture becomes more complex.

Figure 3.17 shows the Doppler spectrum width (curve 1) and shift (curve 2) of the spectral maximum depending on the angle $\Delta\varphi$ between the propagation directions of windsea waves and swell (stationary platform, $V = 0$). The calculations were carried out for two-mode roughness – fully developed waves plus swell. It was assumed that windsea waves ($U = 10$ m s^{-1}) move towards the radar, and the ripples with the dominant wavelength $\Lambda_m = 150$ m are nearly of the maximum height ($H = 2.2$ m). As we see, the dependence of the shift on $\Delta\varphi$ is far weaker than the spectrum breadth.

The shift azimuthal dependences are shown in Figure 3.18: curve 1 indicates the developed windsea ($U = 10$ m s^{-1}); curve 2 stands for both developed windsea and swell ($\Lambda_m = 150$ m, $H = 2.2$ m, $\Delta\varphi = 20°$)); and curve 3 describes the same at $\Delta\varphi = 140°$. It can be seen that the wind wave propagation direction is indicated with an insignificant error.

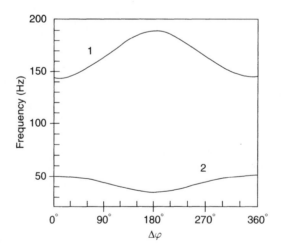

Figure 3.17 Dependence of the Doppler spectrum width (curve1) and shift (curve 2) on the angle $\Delta\varphi$ between the propagation directions of fully developed wind roughness and swell (a stationary platform, $V = 0$). It is assumed that wind waves at the wind speed $U = 10$ m s^{-1} run towards a radar, and swell having the dominant wavelength $\Lambda_m = 150$ m has the height close to the maximal one ($H = 2.2$ m). One can see that the dependence of the shift on $\Delta\phi$ is much weaker in comparison with the spectrum width dependence.

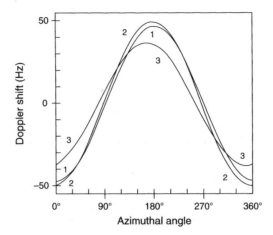

Figure 3.18 Doppler shift azimuthal dependencies: curve 1 – fully developed windsea $U = 10\,\mathrm{m\,s^{-1}}$; curve 2 – the fully developed windsea plus swell ($\Lambda_\mathrm{m} = 150\,\mathrm{m}$, $H = 2.2\,\mathrm{m}$, $\Delta\phi = 20°$); curve 3 – same as 2 but for $\Delta\phi = 140°$. One can see that the travel direction of wind waves can be determined with rather small error.

Comparing the calculation results shown in Figures 3.18–3.21, we conclude that at the incidence angle $\theta_0 = 10°$, the Doppler shift in mixed type roughness is mainly defined by its wind component. This is quite explainable, as wind waves are much more steep than swell and, therefore, the chances of surface elements that would specularly reflect the incident ray are higher specifically for the windsea. Evidently, the predominant influence of the wind waves on the Doppler shift compared with swell grows as the incidence angle increases.

However with the increase in θ_0, first, the intensity of the quasi-specular scatter decreases (see Section 3.2) and, second, as the reflected signal approaches the resonant scatter area, it acquires and later augments the Bragg component. It is exactly the Doppler shift, which depends much on the scatter mechanism, which allows us to estimate the presence level of the Bragg component in the reflected signal. The quasi-specular scattering the Doppler shift is defined by the large wave phase velocities, whereas the Bragg component shift is governed by the phase velocity of a much slower ripple resonant wave. Thus, the gravity centre of the overall signal Doppler spectrum largely depends on the proportion of quasi-specular and Bragg components in the radar signal.

Figure 3.19 depicts the results calculated for the Doppler spectrum of the signal backscattered at the incident angle $\theta_0 = 23°$ (Kanevsky and Karaev 1996b). The calculation has been carried out for $\lambda = 3\,\mathrm{cm}$ and fully developed windsea at the wind speed $U = 6\,\mathrm{m\,s^{-1}}$, $\varphi_0 = 180°$ (upwind radar look direction). The dotted line indicates partial spectra: (1) the Bragg spectrum and (2) the quasi-specular one. As we see, the Bragg component is prevailing here and, therefore, the gravity centre of the Doppler spectrum is near 40 Hz of the resonance ripples frequency, slightly exceeding it.

The situation drastically changes with the appearance on the surface of slick caused by surfactant film, levelling out the resonant ripples. Figure 3.20 displays such a Doppler

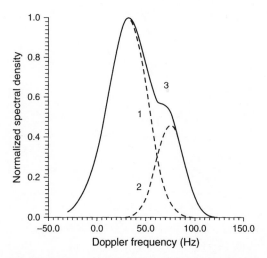

Figure 3.19 The calculated Doppler spectrum of the signal backscattered at the incident angle $\theta_0 = 23°$ [Kanevsky and Karaev 1996b]. The calculations were carried out for $\lambda = 3$ cm and fully developed windsea at wind speed $U = 6$ m s^{-1}, $\phi_0 = 180°$ (upwind radar look direction). Dotted lines are the partial spectra corresponding to resonant (1) and quasi-specular (2) scattering mechanisms. We see in this case the Bragg component is prevalent, therefore the Doppler spectrum gravity point is a little bit over the resonance ripple frequency value 40 Hz.

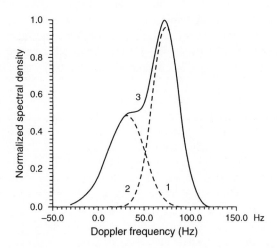

Figure 3.20 The Doppler spectrum calculated in the assumption that the ripple intensity in the slick decreased by 4 dB. Here the quasi-specular component is prevalent, that is why the spectrum gravity point moved to the considerably higher frequency area.

spectrum calculated on the assumption that the ripples intensity in slick area decreased by 4 dB. The quasi-specular component is predominant here, and therefore the Doppler spectrum gravity centre (Doppler centroid) has moved to the much higher frequency area. Figure 3.21 displays the resulting spectra from Figures 3.22 and 3.23.

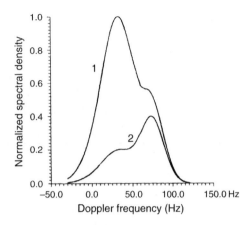

Figure 3.21 Resulting spectra given in Figures 3.19 and 3.20.

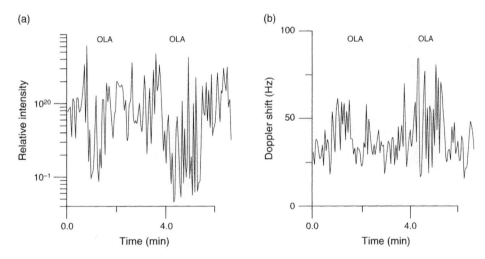

Figure 3.22 The results of synchronous changes in intensity (a) and Doppler shift (b) of backscattered radar signal at the slick passage through radar antenna pattern. Parallel to the signal intensity drop with the transition from outside the slick to inside it the Doppler shift grows from 30 to 35 Hz outside the slicks to 60–70 Hz inside the slick [Kanevsky et al. 1997].

The experiment (Kanevsky et al. 1997) demonstrated a considerably increasing Doppler shift of the signal frequency when returned from slick. The measurements were carried out from the stationary platform in the Black Sea with the use of Doppler radar, working at 2.56 cm microwaves. The incidence angle in the experiments was 18°, the wind speed was about 5 m s^{-1}, and the radar look direction differed from upwind one by 20°. Two artificial slicks of oleyl alcohol (OLA) were deployed on the sea surface.

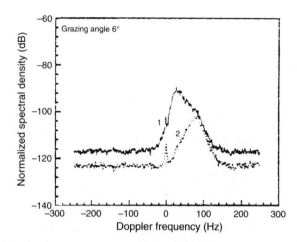

Figure 3.23 Doppler spectra at vertical (1) and horizontal (2) polarizations, $\lambda = 3.2$ cm, depression angle $\psi_0 = 6°$ [Lee et al. 1995a].

The experiment results are presented in Figure 3.22a and b, showing synchronous changes in (a) intensity and (b) Doppler centroid shift of backscattered radar signal at slicks passing through radar antenna pattern. Simultaneously with signal intensity plummeting that accompanies its transition from the area outside slick to the one inside slick, the Doppler centroid frequency increases from 30 to 35 Hz outside the slicks to $60 - 70$ Hz inside the slick. On comparing Figures 3.21 and 3.22 one can conclude that the observed Doppler shifts are in good agreement with theory. Probably, in the case of space-borne radar the effect considered will not be so appreciable. Nevertheless, one can expect this effect to be detectable in as much as the Doppler centroid position can be estimated with high accuracy; in particular, for SEASAT SAR 1.4 Hz accuracy was reached (Li et al. 1982).

As well known, oil slicks may be confused on radar imagery with areas of low wind and other low backscatter features such as cold and freshwater masses (Holt 2004). The adduced effect can be considered as an additional means of confirming the presence of slicks on the sea surface over and above any decrease observed in radar backscatter.

3.4   BACKSCATTERING AT LOW GRAZING ANGLES

Backscattering at low grazing angles plays a specific role in the scattering of electromagnetic microwaves by the rough sea surface. Apart from its application value, this method is of great interest due to the large discrepancy between the resonance (Bragg) scatter theory and the experimental data. (In Figure 3.8, we already mentioned that the backscattering cross section values for horizontally polarized field at the low grazing angles $\psi_0 \leq 10 - 15°$ significantly exceed the resonance backscattering theoretical data.)

The discrepancies between the theory and practical studies at low grazing angles have been pointed out and discussed in a number of works, from the 1960s to the 1970s (Pidgeon 1968, Valenzuela and Laing 1970, Kalmykov and Pustovoitenko 1976])up to this time. These are the main points of discrepancy.

1. The polarization ratio, i.e. the ratio of the backscattering cross-section at horizontal polarization σ_{HH} to that at vertical polarization σ_{VV}, is often over the unity,

$$\chi_{observed} = \left(\frac{\sigma_{HH}}{\sigma_{VV}} \right)_{observed} > 1 \qquad (3.84)$$

   while the Bragg resonant scattering theory predicts very low polarization ratio, $\chi_{Bragg} \ll 1$. Subscripts HH and VV indicate that the transmission and receiving are at the same polarization, horizontal or vertical, respectively. The "super events" frequency (i.e., backscattering events where the polarization ratio exceeds unity) is clearly dependent on the near-surface wind speed (Lee et al. 1996). A characteristic feature is that in the microwave range super events are usually manifested in radar returns, resembling spikes.
2. The Doppler spectra of backscattered fields at opposite polarizations differ a lot. The location of the spectral peak and the spectrum shape at vertical polarization are generally in good accordance with the resonance backscatter theory, which does not hold true for the horizontal polarization field. In the latter case at upwind probing the Doppler spectrum turns out to shift conspicuously to the high frequency area (see Figure 3.23); an important feature is the higher the wind speed, the larger the shift, just like in the super events case.

These peculiarities show that on the surface there are scatterers of some kind, which are faster than the resonance ripples and are responsible for the predominant backscatter of the horizontally polarized pulse at low grazing angles. Since in the microwave range the super events are usually accompanied by the spike-like radar returns, and upwind and downwind directions clearly show disparity here (Kropfli and Clifford 1994), we can conclude that the scatterers are the structures emerging at the wave break.

Under enduring wind conditions, wave breaks occur along the overall length of the surface waves (starting with capillary waves) with the frequency that is the higher, the shorter the respective wave is (Phillips 1988). The breaks have a transient nature; the spatial structure of the breaking wave is constantly changing and generating extensively curved and sloping elements. Several works study these elements' patterns and the diffraction of electromagnetic waves on them – a dielectric wedge modelling a sharp wave crest before the wave breaking (Kalmykov and Pustovoitenko 1976, Lyzenga et al. 1983), as well as the structures shaping water surface straight at the point of breaking and after it (Wetzel 1986).

Lee et al. (1998) analyse the results of scattering experiments from breaking gravity waves conducted at wave tank facility at small grazing angles over the range $4.5 - 11°$. As Lee et al. (1998) remark, in a vigorously breaking wave a sharp crest lives for a short time and impacts the backscatter only at the early stage of wave breaking. The main contribution to the non-Bragg scatter (about 80–90%) is made by the evolving broken crest (or broken wave surface), which is composed of a disordered mass of water, foam and bubbles.

The authors associate the predominance of HH scatter with multi-path effects, when an electromagnetic wave is successively reflected from different surface elements before it is backscattered. In this reflection, the proceeded VV scatter can be largely weakened due to Brewster damping (see Figure 3.3). Wave breaking stops after itself post-break small-scale scatterers on the surface, which are generated by fully breaking gravity waves, from which the Bragg scatter occurs. The prominent role of multi-path effects in backscatter at low grazing angles was also pinpointed in the field experiment (Sletten 1998).

Sletten et al. (2003) highlight the results of very comprehensive radar scattering experiments carried out at a wave tank. Spilling and plunging breakers with a water wavelength of approximately 80 cm were generated and then imaged with a high-speed camera in conjunction with a laser sheet. Simultaneously, the radar backscattering generated by the breakers was measured by an ultra-wideband dual-polarized X-band radar with a range resolution of approximately 4 cm. The nominal grazing angle was 12°. Both the camera and the radar were mounted on a moving instrument carriage that followed the waves throughout their evolution, allowing both sensors to record an uninterrupted record of the breakers space/time evolution. Unlike Lee et al. (1998), the results of this experiment indicate that for the spilling breaker with an up-wave look direction, the crest bulge occurring during the initial stage of breaking is responsible for over 90% of the HH backscatter. The Doppler velocity of the HH backscattering matches the velocity of the crest. In the opinion of the authors (Sletten et al. 2003), the basic HH model in case of spilling breaker is potentially quite simple, as a discrete scatterer (for instance, a cylinder) moving at the phase velocity of the underlying wave exhibits these same characteristics. However, the results indicate that the plunger will require a more complicated scattering model, as more features with a wider range of velocities contribute to its backscatter for both polarizations.

Most works dedicated to microwave scattering on the wave breaking talk about large steep-sloped waves with white caps, when breaking is accompanied by spilling, foam and splutter. The broad array of the scattering mechanisms introduced above work exactly in such breakings. However, the breakings of such magnitude only occur at fairly high wind, while polarization differences at low grazing angles scatter take place in calmer sea as well.

A number of works (see Bulatov et al. (2004) and references therein) mention sharp-crested waves of mesoscale spectrum ("mesowaves") with a characteristic wavelength 30 – 50 cm and height 10 – 20 cm. These waves break without foam and spatter due to their small height; hence the events are not as conspicuous as with white-capped large waves. The works emphasize the important role the mesowaves play in the microwave scatter at low grazing angles.

On the field experiment data (Melief et al. 2006) carried out at a wind speed about 6 m s^{-1}, the authors link sea spikes to breaking of waves with lengths 0.5 – 1.5 m.

According to the results of numerical calculations, performed in Zavorotny and Voronovich (1998), it is possible to achieve good accordance with the experiment outcome at low wind, if alongside free ripples induced by wind and moving orbitally in the field of large ocean waves we take into account also bound (parasitic) capillary waves generated at the front face of gravity waves in the vicinity of their crests. These bound waves velocity approaches that of the gravity wave generating them; this enables associating them with "fast scatterers".

At the same time vertically polarized field is generally scattered in conformity with the Bragg resonant theory, the impact of fast scatterers is relatively small here and the Doppler spectrum of backscattered electromagnetic field at vertical polarization can employ the pattern given in Trizna (1985). As the grazing angle increases, the scattering at both polarizations changes to the Bragg one.

The relative contribution of different backscattering mechanisms depending on polarization, grazing angle and environmental conditions can be studied with the signal processing method suggested in Lee et al. (1995b), according to which the lineshapes of the backscattered microwave power spectra are being decomposed into physically meaningful basis functions which are Gaussian, Lorentzian or Voigtian. Each of these basis functions (spectral lineshapes) corresponds to a concrete backscatter mechanism.

Gaussian lineshape
Within the framework of the resonant (Bragg) scattering model, the backscattered microwave signal Doppler spectrum has a Gaussian profile if the orbital velocities on the water surface have the normal distribution (see Chapter 4). Therefore, it would be natural to assume that the Bragg scattering component corresponds to a Gaussian lineshape in the backscattered signal Doppler spectrum:

$$\phi_G(f) = \frac{1}{\Delta f_G \sqrt{\pi}} \, \exp\left[-\frac{(f - f_G)^2}{(\Delta f_G)^2}\right] \tag{3.85}$$

where f is the current frequency, f_G is the frequency of the spectral maximum and Δf_G is the half-width of the spectral line. Using the relation

$$f = \frac{2v}{\lambda} \, \cos \psi_0 \tag{3.86}$$

where λ is the electromagnetic wavelength, we introduce a new variable, the scatterer velocity v, so that

$$\phi_G(v) = \frac{1}{\Delta v_G \sqrt{\pi}} \, \exp\left[-\frac{(v - v_G)^2}{(\Delta v_G)^2}\right] \tag{3.87}$$

Lorentz lineshape
A Lorentz spectral lineshape corresponds to the case where a uniformly moving (i.e. with a steady speed) emitter (or scatterer) exists during a finite time called the "lifetime". This lineshape can reasonably be related to breaking of the dominant wave or wave close to the dominant one, which moves to a certain phase velocity. (Note that the phase velocity of the dominant wave significantly exceeds the phase velocity of the resonant ripples.) Most likely, a spike of the backscattered signal and super event accompanying it are caused by the breaking of exactly dominant (or close to that) wave, when the wave element responsible for the predominant scattering of horizontally polarized radiation appears and later vanishes.

With Eqn. (3.86), the expression for the Lorentz line is written as

$$\phi_L(v) = \frac{1}{\pi} \frac{\Delta v_L}{(v - v_L)^2 + (\Delta v_L)^2} \tag{3.88}$$

where v_L corresponds to the maximum of $\phi_L(v)$ and Δv_L is the half-width at half-maximum of the Lorentzian process.

Voigt lineshape
This lineshape, well-known in astrophysics, is the convolution of Gaussian and Lorentzian profiles

$$\phi_V(v) = \frac{a}{\Delta v'_G \pi^{3/2}} \int\limits_{-\infty}^{+\infty} \frac{\exp(-y^2)\, dy}{\left[(v - v_V)/\Delta v'_G - y\right]^2 + a^2} \tag{3.89}$$

The quantity v_V corresponds to the peak in a Voigt spectrum, $\Delta v'_G$ is the e-folding half-width of the Gaussian line, the Voigt parameter $a = \Delta v'_L/\Delta v'_G$ is the ratio of the Lorentz and Gaussian line half-widths. The primes stand for the parameters of the Lorentz and Gaussian lines in convolution (3.89) in contrast to the similar parameters in Eqns (3.87) and (3.88). Physically, a Voigt line profile corresponds to scatterers with finite lifetimes and normal velocity distributions. It is evident that Δv_G and $\Delta v'_G$ are different, since these values describe the velocity distributions of Bragg and fast scatterers, respectively. A Voigt lineshape should probably be related to breaking of the continuum of smaller waves having different velocities (both proper phase velocities and orbital velocities in the dominant wave field).

The Doppler spectrum $D(v)$ can be decomposed as follows:

$$D(v) = \sum_{i=G,L,V} C_i \phi_i(v) \tag{3.90}$$

In this case, the coefficients C_i describe contributions of each mechanism in the scattered field. The lack of rigor of this approach, related to the fact that the used set of lineshapes $\phi_i(v)$ is not a complete system of orthogonal functions, is compensated by a priori allowing for the models of scattering mechanisms.

Figure 3.24a and b (quoted from Lee et al. 1995b) illustrates this method with examples. The Doppler spectra of backscattered signal of microwave frequency of 9.23 GHz were obtained for an upwind-look configuration at a grazing angle of 10°. To make another computation of the Doppler frequency for the scatterer velocity, it is necessary to include the speed of the boat where the sensing was performed. As the article supplies no information on the carrier speed, we did not make any such computations.

Figure 3.25a and b gives the same data for $\lambda = 3.2$ cm, obtained in the experiment (Kanevsky et al. 2001). One can see that in both cases (i.e. in Figure 3.24 and in Figure 3.25) the spectra are represented by identical sets of base functions $\phi_i(v)$.

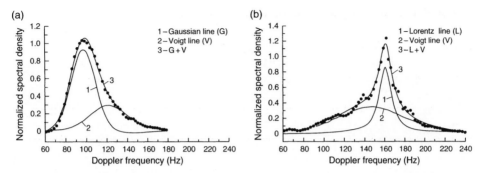

Figure 3.24 Doppler spectra of signals with vertical (a) and horizontal (b) polarizations; the wavelength $\lambda = 3.2\,\text{cm}$, the depression angle $\psi_0 = 10°$, the wind speed $U = 9\,\text{m s}^{-1}$. The measurements were carried out from the moving boat [Lee et al. 1995b].

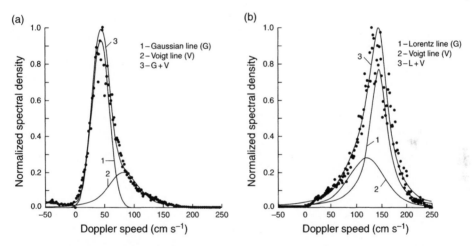

Figure 3.25 Doppler spectra of signals with vertical (a) and horizontal (b) polarizations. The wavelength $\lambda = 3.2\,\text{cm}$, the depression angle $\psi_0 = 5°$, the wind speed $U = 7\,\text{m s}^{-1}$; the frequencies are re-calculated into the effective scatterer speed [Kanevsky et al. 2001].

In accordance with Lee et al. (1995b), we can draw the following conclusions from the results of the Doppler spectra processing effected by decomposing them into physically meaningful basis functions.

1. Scattering from the free Bragg waves characterized by a Gaussian distribution in scatterer speeds, resulting in a low-frequency Gaussian component of the Doppler spectrum.
2. Scattering from the sporadically appearing, short lifetime, fast-moving, "single-speed" (at the phase speed of the gravity wave), facets of a breaking wave, characterized by a Lorentzian component of the spectrum at high frequency.

3. Scattering from fast to intermediate speed breaking shorter gravity waves of different wavelengths (and thus with a spread in values of phase speed), and/or short lifetime-bound Bragg waves (i.e. short relative to the lifetime of free Bragg waves), characterized by a convolution of the Gaussian and Lorentzian processes, resulting in a Voigtian profile in the spectrum.

Summing up this section, we conclude that although numerous studies of microwave scattering at low grazing angles have elicited its peculiarities, the picture of the phenomenon is not yet complete.

– 4 –

Microwave Doppler spectrum at moderate incidence angles

The most solid criteria of the scattering theory practicability might be the proximity of the backscattered field theoretical and experimental Doppler spectra. Recall that the first and essentially the drastic step to the scatter mechanism comprehension was exactly based on the HF signal Doppler spectrum analysis.

This chapter deals with the Doppler spectrum of microwave field backscattered (i.e. scattered towards radar) by rough sea surface at moderate incidence angles, where the resonant scattering takes place (Fuks 1974, Plant and Keller 1990). We have devoted a special chapter to the Doppler spectrum to emphasize the importance of this feature of the scattering phenomenon. Besides, this importance is also due to the wide use of the Doppler spectrum width and spectral maximum location in various aspects of radar remote sensing of the ocean.

Figure 4.1 presents the surface illumination scheme; the central ray is directed to the point x_0, y_0, on the plane XY. The illuminated surface spot linear values D_x, D_y along the X, Y axes are determined by the angle width antenna pattern δ_x, δ_y in the vertical and azimuthal planes, respectively. The slant range coordinates R_0, R_1, R_2 lie in the plane XY, and R is in the point on the random surface $\zeta(x', y')$, formed by a large ocean wave. The depression angle ψ_0 ranges within the feasibility limits of resonance scatter. Evidently,

$$R_2^2 = R_1^2 + (y - y_0)^2, \quad R_1^2 = R_0^2 \, \sin^2 \psi_0 + (R_0 \cos \psi_0 + x - x_0)^2 \quad (4.1)$$

which yields

$$R_2^2 = R_0^2 + 2R_0(x - x_0) \cos \psi_0 + (x - x_0)^2 + (y - y_0)^2 \quad (4.2)$$

or

$$R_2 \approx R_0 + (x - x_0) \cos \psi_0 + \frac{1}{2R_0} \left[(x - x_0)^2 + (y - y_0)^2 \right] \quad (4.3)$$

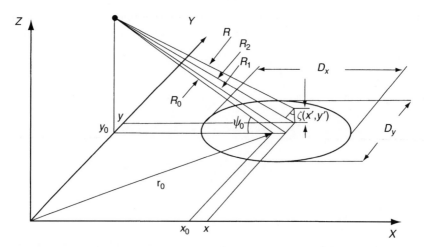

Figure 4.1 The surface irradiation scheme (on the calculation of the Doppler spectrum).

Supposing $R >> \sigma_{\zeta}^2/\lambda$ is satisfied, we write

$$R \approx R_2 - \zeta \sin \psi_0 \qquad (4.4)$$

and then

$$R \approx R_0 + (x - x_0) \cos \psi_0 - \zeta \sin \psi_0 + \frac{1}{2R_0} \left[(x - x_0)^2 + (y - y_0)^2 \right] \qquad (4.5)$$

In accordance with the composite theory based on the two-scale model of the sea surface the backscattered microwave field source proper is small gravity–capillary waves (ripples). Large-scale waves are manifested by the signal amplitude modulation which is due to, first, the different local depression angles and, second, the ripples amplitude hydrodynamic modulation along the large wave. Besides, temporal shifts in the arbitrary range R caused by large-scale waves induce backscattered field frequency shifts.

Hence the expression for the backscattered electromagnetic field complex amplitude in the perturbation theory first approximation is defined as

$$a\left(\vec{r}_0, t\right) = \frac{2k^2}{\sqrt{\pi}} \int d\vec{r}\, \phi\left(\vec{r} - \vec{r}_0\right) m(\vec{r}, t) \xi(\vec{r}, t) \exp\left[2ikR(\vec{r}, t)\right] \qquad (4.6)$$

Here $\phi(\vec{r} - \vec{r}_0)$ is a non-zero function only inside the illuminated area, $\xi(\vec{r}_1, t)$ is the function describing "standard" ripples, i.e. with constant mean characteristics along the large wave, and $m(\vec{r}, t)$ is a function for the backscattered signal amplitude modulation (geometric and hydrodynamic) by large-scale waves. Besides, as in Eqn. (4.6) integration is not over the surface ζ but over the surface $z = 0$; function $m(\vec{r}, t)$ includes also the multiplier $1/J$, where $J = \left[\left(1 - n_x^2\right)\left(1 - n_y^2\right)\right]^{1/2}$ is the Jacobian of the transformation;

and $n_x = -\partial\zeta/\partial x$ and $n_y = -\partial\zeta/\partial y$ are the projections on plane $z = 0$ of the local normal to the surface ζ. However, due to small slopes of large-scale waves at moderate winds, the variable J is close to unity.

Note that the complex amplitude $a(\vec{r}_0, t)$ is normalized, so as its modulus squared and averaged over the surface ζ realizations gives an average backscattering cross section (3.14).

Using Eqn. (4.5) for the current range R from radar to the point on the surface $\zeta(\vec{r}_1, t)$, the reflected signal complex amplitude correlation function is expressed as

$$
\begin{aligned}
B(\tau) &= \langle a(t)a^*(t+\tau)\rangle \\
&= \frac{4k^4}{\pi}\int\int d\vec{r}'\,d\vec{r}''\,\phi\left(\vec{r}'-\vec{r}_0\right)\phi\left(\vec{r}''-\vec{r}_0\right) \\
&\quad \times \left\langle m\left(\vec{r}',t\right)m^*\left(\vec{r}'',t+\tau\right)\xi\left(\vec{r}',t\right)\xi\left(\vec{r}'',t+\tau\right) \right. \\
&\quad \left. \times \exp\left\{2ik\left[\begin{array}{l}(x'-x'')\cos\psi_0 - \left[\zeta\left(\vec{r}',t\right) - \zeta\left(\vec{r}'',t+\tau\right)\left(\vec{r}'',t+\tau\right)\right]\sin\psi_0 \\ +\dfrac{1}{2R_0}\left[(x'-x_0)^2-(x''-x_0)^2+(y'-y_0)^2-(y''-y_0)^2\right]\end{array}\right]\right\}\right\rangle
\end{aligned}
$$

$$(4.7)$$

We change over to new variables:

$$
\begin{aligned}
\vec{r}'-\vec{r}'' &= \vec{\rho}, & \vec{r}' &= \vec{r}+\vec{\rho}/2 \\
\vec{r}'+\vec{r}'' &= 2\vec{r}, & \vec{r}'' &= \vec{r}-\vec{\rho}/2
\end{aligned}
$$

$$(4.8)$$

and with regard to the statistic independence of standard ripples and large-scale waves average separately over the ripples realizations:

$$
B(\tau) = \frac{4k^2}{\pi}\int d\vec{r}\,\Phi\left(\vec{r}-\vec{r}_0\right)\left\langle |m(\vec{r},t)|^2 \exp\left[-2ik\tau\frac{\partial\zeta}{\partial t}(\vec{r},t)\sin\psi_0\right]\int d\vec{\rho}\,B_\xi(\vec{\rho},\tau)\right.
$$
$$
\left. \times \exp\left\{2ik\left[\begin{array}{l}\left(\cos\psi_0 - \dfrac{\partial\zeta}{\partial x}(\vec{r},t)\sin\psi_0 + \dfrac{1}{R_0}(x-x_0)\right)\rho_x \\ + \left(\dfrac{1}{R_0}(y-y_0) - \dfrac{\partial\zeta}{\partial y}(\vec{r},t)\sin\psi_0\right)\rho_y\end{array}\right]\right\}\right\rangle,
$$

$$(4.9)$$

where $\Phi = \phi^2$ and B_ξ represents the standard ripples correlation function. When expressing Eqn. (4.9) we made allowance for the weak fluctuation in function ϕ and large wave characteristics on the spatial and temporal scale of ripples correlation, where $B_\xi(\vec{\rho},\tau)$ decreases significantly, therefore:

$$
\phi\left(\vec{r}'-\vec{r}_0\right) \approx \phi\left(\vec{r}''-\vec{r}_0\right) \approx \phi\left(\vec{r}-\vec{r}_0\right)
$$

$$(4.10)$$

$$m\left(\vec{r}',t\right) \approx m\left(\vec{r}'',t+\tau\right) \approx m\left(\vec{r},t\right) \tag{4.11}$$

$$\zeta\left(\vec{r}',t\right) - \zeta\left(\vec{r}'',t+\tau\right) \approx \frac{\partial\zeta}{\partial x}\left(\vec{r},t\right)\rho_x + \frac{\partial\zeta}{\partial y}\left(\vec{r},t\right)\rho_y + \frac{\partial\zeta}{\partial t}\left(\vec{r},t\right)\tau \tag{4.12}$$

In accordance with the Wiener–Khinchin theorem, $B_\xi(\vec{\rho},\tau)$ is described as

$$B_\xi(\vec{\rho},\tau) = \int\int d\vec{\rho}\; d\omega \Psi_\xi(\vec{\kappa},\omega)\exp\left[i(\vec{\kappa}\,\vec{\rho} - \omega\tau)\right] \tag{4.13}$$

where $\Psi_\xi(\vec{\kappa},\omega)$ stands for the spatio-temporal ripples spectrum. Here and elsewhere, we shall omit the term "standard" to describe ripples, bearing in mind however that we mean a statistically homogeneous field, while real ripples amplitude modulation is calculated with the multiplier $m(\vec{r},t)$.

Multiplying both the parts of Eqn. (4.13) by $\exp(-i\vec{\kappa}\,\vec{\rho})$ and integrating in the infinite limits over $\vec{\rho}$, we get

$$\int d\vec{\rho}\, B_\xi(\vec{\rho},\tau)\exp(-i\vec{\kappa}\,\vec{\rho}) = 4\pi^2 \int d\omega \Psi_\xi(\vec{\kappa},\omega)\exp(-i\omega\tau) \tag{4.14}$$

Inserting Eqn. (4.14) into Eqn. (4.9) yields

$$\begin{aligned}
B(\tau) = 16\pi k^4 &\int d\vec{r}\; \Phi\left(\vec{r} - \vec{r}_0\right) \\
&\times \left\langle \left|m(\vec{r},t)\right|^2 \int d\omega \Psi_\xi\left(\vec{\kappa}_{\text{res}},\omega\right)\exp\left[-i\left(\omega + 2k\frac{\partial\zeta}{\partial t}\sin\psi_0\right)\tau\right]\right\rangle
\end{aligned} \tag{4.15}$$

where

$$\begin{aligned}
\vec{\kappa}_{\text{res}}(\vec{r},t) = \Big\{ &-2k\left[\cos\psi_0 - \frac{\partial\zeta}{\partial x}(\vec{r},t)\sin\psi_0 + \frac{1}{R_0}(x - x_0)\right]; \\
&-2k\left[\frac{1}{R_0}(y - y_0) - \frac{\partial\zeta}{\partial y}(\vec{r},t)\sin\psi_0\right]\Big\}
\end{aligned} \tag{4.16}$$

is a local resonance wave number, which varies with the slopes $\partial\zeta/\partial x$ and $\partial\zeta/\partial y$ of the large-scale wave surface in the arbitrary point $\vec{r}$ as well as with the point location against the illuminated area centre $\vec{r}_0$. It is to point out that the vector $\vec{\kappa}_{\text{res}}$, computed from expression (4.16), does not differ from the one that occurs from formulas (3.36a) and (3.36b), if we take into account that $\alpha_x \approx \partial\zeta/\partial x \ll 1$, $\alpha_y \approx \partial\zeta/\partial y \ll 1$, and set $x = x_0, y = y_0$.

Due to the smallness of slopes and values $(D_x, D_y)/R$ we show that

$$\vec{\kappa}_{\text{res}} = \{-2k\cos\psi_0; 0\} \tag{4.17}$$

Since centimetre ripples move orbitally on the large wave, their frequency, i.e. the ripple spectrum resonance component shifts and becomes equal to $\Omega_{\text{res}} + \vec{\kappa}_{\text{res}} \vec{v}_{\text{orb}}$, where Ω_{res} represents inherent frequency and $\vec{v}_{\text{orb}}$ is the orbital velocity. Let there be two counter-propagating waves on the surface, then

$$
\begin{aligned}
\Psi_\xi\left(\vec{\kappa}_{\text{res}}, \omega\right) = \hat{W}_\xi^+\left(\vec{\kappa}_{\text{res}}\right) \delta\left(\omega - \left(\Omega_{\text{res}} + \vec{\kappa}_{\text{res}} \vec{v}_{\text{orb}}\right)\right) \\
+ \hat{W}_\xi^-\left(\vec{\kappa}_{\text{res}}\right) \delta\left(\omega + \left(\Omega_{\text{res}} - \vec{\kappa}_{\text{res}} \vec{v}_{\text{orb}}\right)\right)
\end{aligned}
\tag{4.18}
$$

The summands on the right-hand side of Eqn. (4.18) stand for the counter-propagating waves as they have temporal frequencies of different signs at one and the same wave vector. Remember that $\hat{W}_\xi^-\left(\vec{\kappa}_{\text{res}}\right) = \hat{W}_\xi^+\left(-\vec{\kappa}_{\text{res}}\right)$; therefore,

$$
\begin{aligned}
\Psi_\xi\left(\vec{\kappa}_{\text{res}}, \omega\right) = \hat{W}_\xi^+\left(\vec{\kappa}_{\text{res}}\right) \delta\left(\omega - \left(\Omega_{\text{res}} + \vec{\kappa}_{\text{res}} \vec{v}_{\text{orb}}\right)\right) \\
+ \hat{W}_\xi^-\left(\vec{\kappa}_{\text{res}}\right) \delta\left(\omega + \left(\Omega_{\text{res}} - \vec{\kappa}_{\text{res}} \vec{v}_{\text{orb}}\right)\right) \\
= \frac{1}{2} W_\xi\left(\vec{\kappa}_{\text{res}}\right) \delta\left(\omega - \left(\Omega_{\text{res}} + \vec{\kappa}_{\text{res}} \vec{v}_{\text{orb}}\right)\right) \\
+ \frac{1}{2} W_\xi\left(-\vec{\kappa}_{\text{res}}\right) \delta\left(\omega + \left(\Omega_{\text{res}} - \vec{\kappa}_{\text{res}} \vec{v}_{\text{orb}}\right)\right)
\end{aligned}
\tag{4.19}
$$

We substitute Eqn. (4.19) into Eqn. (4.15) and integrate over ω:

$$
\begin{aligned}
B(\tau) = \int d\vec{r} \, \Phi(\vec{r} - \vec{r}_0) \langle \sigma_0^+ \exp[-i(\Omega_{\text{res}} + 2kv_{\text{rad}})\tau] \rangle \\
+ \int d\vec{r} \, \Phi(\vec{r} - \vec{r}_0) \langle \sigma_0^- \exp[i(\Omega_{\text{res}} - 2kv_{\text{rad}})\tau] \rangle
\end{aligned}
\tag{4.20}
$$

Here $\sigma_0^\pm = 8\pi\kappa^4 |m(\vec{r}, t)|^2 W_\xi(\pm\kappa_{\text{res}})$ are the local radar cross sections, corresponding to the two waves, and v_{rad} is a radial component of orbital velocity considered positive when directed towards the radar. Multiplier $|m|^2$ unlike Eqn. (3.11) includes both geometric and hydrodynamic modulations of the reflected signal. Expression (4.20) takes into account that

$$
\vec{\kappa}_{\text{res}} \vec{v}_{\text{orb}} + 2k \frac{\partial \zeta}{\partial t} \sin \psi_0 = -2 \, \vec{k} \vec{v}_{\text{orb}} = 2kv_{\text{rad}}
\tag{4.21}
$$

Generally, the normalized radar cross section can be represented in the form

$$
\sigma_0^\pm = \langle (\sigma_0^\pm) + \delta\sigma_{0,\text{lin}}^\pm \rangle + \delta\sigma_{0,\text{nlin}}^\pm
\tag{4.22}
$$

where $\delta\sigma_{0,\text{lin}}^{\pm}$ and $\delta\sigma_{0,\text{nlin}}^{\pm}$ are the linear and nonlinear parts of fluctuations of $\sigma_0^{\pm}$ respectively. The linear part considered in Section 3.2.2 is related to the surface $\zeta(\vec{r},t)$ elevations through the linear MTF:

$$\delta\sigma_{0,\text{lin}}^{\pm} = \langle\sigma_0^{\pm}\rangle \int d\vec{\kappa}\ T(\vec{\kappa})A_\zeta(\vec{\kappa})\exp\left[i\left(\vec{\kappa}\vec{r} - \Omega(\kappa)\right)t\right] + \text{c.c.} \tag{4.23}$$

Here $T(\vec{\kappa}) = T_{\text{tilt}}(\vec{\kappa}) + T_{\text{hydr}}(\vec{\kappa})$ is the linear MTF describing two kinds of modulation (tilt and hydrodynamic) and $A_\zeta(\vec{\kappa})$ is the spectral amplitude in the decomposition

$$\zeta(\vec{r},t) = \int d\vec{\kappa}\ A_\zeta(\vec{\kappa})\exp\left[i\left(\vec{\kappa}\vec{r} - \Omega(\kappa)\right)t\right] + \text{c.c.} \tag{4.24}$$

We restrict ourselves to the linear representation and write

$$\langle\sigma_0^{\pm}\exp(-2ikv_{\text{rad}}\tau)\rangle = \langle\sigma_0^{\pm}\rangle\langle\exp(-2ikv_{\text{rad}}\tau)\rangle + \langle\delta\sigma_{0,\text{lin}}^{\pm}\exp(-2ikv_{\text{rad}}\tau)\rangle \tag{4.25}$$

Note that v_{rad} is just like $\delta\sigma_{0,\text{lin}}^{\pm}$ the linear function of $\zeta(\vec{r},t)$, therefore

$$v_{\text{rad}}(\vec{r},t) = \int d\vec{\kappa}\ T_{v_{\text{rad}}}(\kappa)A_\zeta(\vec{\kappa})\exp\left[i\left(\vec{\kappa}\vec{r} - \Omega(\kappa)\right)t\right] + \text{c.c.} \tag{4.26}$$

where (see Hasselmann and Hasselmann, 1991)

$$T_{v_{\text{rad}}}(\vec{\kappa}) = -\Omega(\kappa)\left(\frac{\kappa_x}{\kappa}\cos\psi_0 + i\sin\psi_0\right) \tag{4.27}$$

is the MTF of v_{rad}. (The first term in Eqn. (4.27) arises from the horizontal component of the orbital velocity, while the second term, in quadrature with the first, from the vertical component.)

Thus, both $\sigma_{0,\text{lin}} = \langle\sigma_0\rangle + \delta\sigma_{0,\text{lin}}$ and v_{rad} obey the normal distribution with joint PDF (we will hereafter omit the subscript "lin" to shorten the expression)

$$p(\sigma_0, v_{\text{rad}}) = \frac{1}{2\pi\sigma_{\text{rad}}\sigma_{\delta\sigma_0}\sqrt{1 - r^2_{v,\delta\sigma_0}}}$$

$$\times\exp\left\{-\frac{1}{2\left(1 - r^2_{v,\delta\sigma_0}\right)}\left[\frac{(\delta\sigma_0)^2}{\sigma^2_{\delta\sigma_0}} - 2r_{v,\delta\sigma_0}\frac{v_{\text{rad}}\delta\sigma_0}{\sigma_{\text{rad}}\sigma_{\delta\sigma_0}} + \frac{v^2_{\text{rad}}}{\sigma^2_{\text{rad}}}\right]\right\} \tag{4.28}$$

where σ_{rad} and $\sigma_{\delta\sigma_0}$ are the RMS of v_{rad} and $\delta\sigma_0$, respectively, and

$$r_{v,\delta\sigma_0} = \frac{1}{\sigma_{\text{rad}},\sigma_{\delta\sigma_0}}\langle v_{\text{rad}}\delta\sigma_0\rangle \tag{4.29}$$

is the correlation coefficient of these values.

Equation (4.28) yields

$$\langle \delta\sigma_0 \exp(-2ikv_{\text{rad}}\tau)\rangle = -2ir_{v,\delta\sigma_0}\sigma_{\delta\sigma_0}k\sigma_{\text{rad}}\tau\exp\left(-2k^2\sigma_{\text{rad}}^2\tau\right) \qquad (4.30)$$

Therefore

$$B(\tau) = B_0(\tau) + B_1(\tau) \qquad (4.31)$$

where

$$B_0(\tau) = \frac{1}{\sqrt{2\pi}\sigma_{\text{rad}}}\left\{\int d\vec{r}\,\Phi(\vec{r}-\vec{r}_0)\int dv_{\text{rad}}\langle\sigma_0^+\rangle\exp\left[-\frac{v_{\text{rad}}^2}{2\sigma_{\text{rad}}^2} - i(\Omega_{\text{res}}+2kv_{\text{rad}})\tau\right]\right.$$
$$\left. +\int d\vec{r}\,\Phi(\vec{r}-\vec{r}_0)\int dv_{\text{rad}}\langle\sigma_0^-\rangle\exp\left[-\frac{v_{\text{rad}}^2}{2\sigma_{\text{rad}}^2} + i(\Omega_{\text{res}}-2kv_{\text{rad}})\tau\right]\right\}$$

$$(4.32)$$

and

$$B_1(\tau) = -2iS_0k\sigma_{\text{rad}}\tau\begin{bmatrix}r_{v,\delta\sigma_0}^+\,\sigma_{\delta\sigma_0}^+\exp\left(-2k^2\sigma_{\text{rad}}^2\tau^2 - i\Omega_{\text{res}}\tau\right)\\ +r_{v,\delta\sigma_0}^-\sigma_{\delta\sigma_0}^-\exp\left(-2k^2\sigma_{\text{rad}}^2\tau^2 + i\Omega_{\text{res}}\tau\right)\end{bmatrix} \qquad (4.33)$$

where $S_0 = \int d\vec{r}\,\Phi(\vec{r}-\vec{r}_0)$. Then, integrating Eqn. (4.32) over v_{rad}, we get

$$B_0(\tau) = S_0\left[\langle\sigma_0^+\rangle\exp\left(-2k^2\sigma_{\text{rad}}^2\tau^2 - i\Omega_{\text{res}}\tau\right) + \langle\sigma_0^-\rangle\exp\left(-2k^2\sigma_{\text{rad}}^2\tau^2 + i\Omega_{\text{res}}\tau\right)\right] \quad (4.34)$$

Switching over from the correlation function to the temporal spectrum,

$$G(\omega) = \frac{1}{\pi}\int d\tau\,B(\tau)\exp(i\omega\tau) \qquad (4.35)$$

we obtain

$$G(\omega) = G_0(\omega) + G_1(\omega) \qquad (4.36)$$

$$G_0(\omega) = \frac{S_0}{\sqrt{2\pi}k\sigma_{\text{rad}}}\left\{\langle\sigma_0^+\rangle\exp\left[-\frac{(\omega-\Omega_{\text{res}})^2}{8k^2\sigma_{\text{rad}}^2}\right] + \langle\sigma_0^-\rangle\exp\left[-\frac{(\omega+\Omega_{\text{res}})^2}{8k^2\sigma_{\text{rad}}^2}\right]\right\} \qquad (4.37)$$

$$G_1(\omega) = \frac{S_0}{\sqrt{\pi}k\sigma_{\text{rad}}}\left\{ r_{v,\delta\sigma_0}^+\sigma_{\delta\sigma_0}^+\frac{\omega - \Omega_{\text{res}}}{2\sqrt{2}k\sigma_{\text{rad}}}\exp\left[-\frac{(\omega - \Omega_{\text{res}})^2}{8k^2\sigma_{\text{rad}}^2}\right]\right.$$

$$\left. + r_{v,\delta\sigma_0}^-\sigma_{\delta\sigma_0}^-\frac{\omega - \Omega_{\text{res}}}{2\sqrt{2}k\sigma_{\text{rad}}}\exp\left[-\frac{(\omega + \Omega_{\text{res}})^2}{8k^2\sigma_{\text{rad}}^2}\right]\right\}$$

$$(4.38)$$

The coefficient $1/\pi$, instead of $1/2\pi$, before the integral in Eqn. (4.35) means that we analyse only positive temporal frequencies.

As we see, the basis of spectrum G_0, as in HF diapason, is two lines at frequencies $\pm\Omega_{\text{res}}$; their intensity relation is determined via the ratio $\langle\sigma_0^-\rangle/\langle\sigma_0^+\rangle$, and width via the RMS of orbital velocity radial component.

In our analysis we supposed for simplicity that the average of the orbital velocity radial component is zero, hence the peaks of the Doppler spectrum $G_0(\omega)$ turned out to be spaced apart symmetrically against zero at the distances equal to the frequency of the roughness spectrum resonance (Bragg) component. Under in situ conditions, as we already know, the Doppler spectrum peaks, as a rule, shift as a whole to either direction, governed by a number of factors, such as currents, wind drift and Stokes drift currents (see Chapter 3).

The additionally introduced member $G_1(\omega)$ describes the asymmetry of the spectral lines against $\pm\Omega_{\text{res}}$, caused by the modulation of radar cross section. The fact is that the radar cross section modulation (tilt and hydrodynamic) induces the intensity variance between the signals backscattered by large wave symmetric areas, where the v_{rad} values have identical modulus but opposite signs. Therefore, the Doppler spectrum of the reflected signal features mostly the frequencies corresponding to the large wave area with the more intense backscatter, which accounts for the line asymmetry. The values $r_{v,\delta\sigma_0}^\pm$ occurring in Eqn. (4.38) can be found (see Appendix A) as

$$r_{v,\delta\sigma_0}^\pm = \frac{\langle\sigma_0^\pm\rangle}{\sigma_{\text{rad}}\sigma_{\delta\sigma_0}^\pm}\int d\vec{\kappa}\,\text{Re}\left[T(\vec{\kappa})T_{v_{\text{rad}}}^*(\vec{\kappa})\right]W_\zeta(\vec{\kappa}) \qquad (4.39)$$

where $T(\vec{\kappa})$ and $T_{v_{\text{rad}}}(\vec{\kappa})$ are MTF for σ_0 and v_{rad}, respectively. Strictly speaking, another outcome of the cross-section modulation is also some extra peak shift, besides the one mentioned above.

The Doppler spectrum lineshape can be found more accurately, if we introduce non-linear fluctuations $\delta\sigma_{0,\text{nlin}}$ (see Eqn. (4.22)) in our computation, as it was done in Romeiser and Thompson (2000), which apart from linear fluctuations of σ_0 described by linear MTF, took into account the second-order NRCS fluctuations, i.e. $\delta\sigma_{0,\text{nlin}} \propto \zeta^2$.

Equation (4.37) does not take into account the polarization dependence of the Doppler lineshape, as it is fairly little manifested at moderate incidence angles (unlike low grazing angles, where, as we know, opposite polarization signals employ different scattering mechanisms). At the same time, the comparison of the τ_{cor} values – the correlation time of the reflected field corresponding to the simple expression (4.37) and numerical calculations performed by Romeiser and Thompson (2000), proves their proximity. The conclusion has been suggested by comparing the data obtained from Figure 7 of the cited article and those given by the formula following from (4.34):

$$\tau_{cor} = \frac{\lambda}{2\sqrt{2}\pi\sigma_{rad}} \tag{4.40}$$

if we define τ_{cor} according to the decay degree $1/e$ of the field autocorrelation function or $\tau_{cor}^{(0.5)} = 0.83\tau_{cor}$ at level 0.5.

Let

$$\sigma_{rad} = \left(\sin^2\psi_0 + \cos^2\varphi_0 \cos^2\psi_0 \right)^{1/2} \sigma_{orb} \tag{4.41}$$

where φ_0 is the angle between the radar look direction (horizontal plane projection) and the general direction of the surface wave propagation, σ_{orb} orbital velocity RMS. For developed windsea described by the Pierson–Moskowitz spectrum $\sigma_{orb} = \sigma_{vert} = 6.9 \times 10^{-2}U$, where U stands for the near-surface wind speed (see Eqn. (2.23)). Then

$$\tau_{cor} \approx \frac{1.65\lambda}{\left(\cos^2 \theta_0 + \cos^2 \varphi_0 \sin^2 \theta_0 \right)^{1/2}U} \tag{4.42a}$$

$$\tau_{cor}^{(0.5)} \approx \frac{1.3\lambda}{\left(\cos^2 \theta_0 + \cos^2 \varphi_0 \sin^2 \theta_0 \right)^{1/2}U} \tag{4.42b}$$

As the computation from Eqn. (4.42b) has shown, the discrepancy between the values of $\tau_{cor}^{(0.5)}$ obtained by two different methods varies from 5% to 15% at the same parameter values as in Romeiser and Thompson (2000).

Now we will find the microwave Doppler spectrum linewidth. According to Eqn. (4.37) the width Δf (Hz) at the $1/e$ level is $4\sqrt{2}\sigma_{rad}/\lambda$. Therefore, as function of angle φ_0 Doppler linewidth varies within the interval

$$0.4\frac{U}{\lambda} \sin\psi_0 \le \Delta f \le 0.4\frac{U}{\lambda} \tag{4.43}$$

Obviously, the lines are resolved provided their half-width does not exceed the distance between them, i.e. $\Delta f \le 4f_{res}$.

Formula (4.43) yields spectral linewidth $\Delta f \approx 100$ Hz, provided $\lambda = 3.2$ cm, $U = 8$ m s^{-1} and the survey is along or against the wind. The result is in good agreement with the experimental data shown in Figure 4.2 from Plant and Keller (1990). Note that all spectra in Figure 4.2 are unimodal, because, first, every linewidth is significantly larger than the distance between them, and second, lines that correspond to the surface waves directed along or against wind differ very much in their intensity. The spectrum evolution as a function of azimuth angle φ_0 is displayed in Figure 4.3 (Poulter et al. 1994), which allows us to trace width variations and line intensity relations. This, particularly, indicates

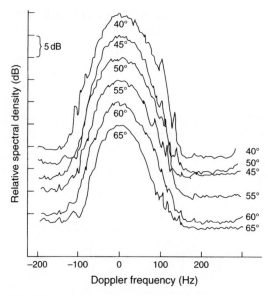

Figure 4.2 The set of Doppler spectra at various incidence angles, wind speed $U = 7 - 9\,\mathrm{m\,s^{-1}}$ and electromagnetic wavelength $\lambda = 3.2\,\mathrm{cm}$ (Plant and Keller 1990).

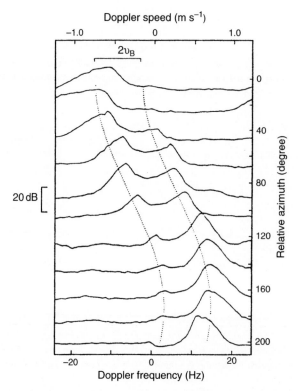

Figure 4.3 Azimuthal dependence of the Doppler spectrum (Poulter et al. 1994).

that the ripple spectrum almost lacks the components with a negative projection on the wind direction – in any case, at azimuth angles of 0° and 180° such lines are not present practically.

A good agreement of theoretical and experimental values of Doppler width is presented in Figure 4.4 which shows experimental data of both Plant and Keller (1990) and Grebenyuk et al. (1994). The results given in Grebenyuk et al. (1994) testify to the high information capacity of Doppler spectrum width for the distant sea roughness diagnostics.

A specific character is displayed by the spectra obtained via narrow-directed antenna in the timespan short enough compared to the large wave period. It is easily understood that these spectra prove to be narrow and non-stationary – they periodically shift in either direction, depending on the value and direction of the orbital velocity in the large wave illuminated area (Figure 4.5; Rozenberg et al. (1973)). Once the function $s(t)$ is written for

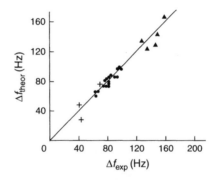

Figure 4.4 Comparison of theoretical and experimental Doppler width values. Here the experimental data of both Plant and Keller (1990) (triangles: $\lambda = 3.2$ cm, pluses: $\lambda = 7$ cm) and Grebenyuk et al. (1994) (circles: $\lambda = 3.2$ cm) are given.

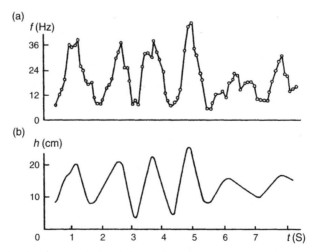

Figure 4.5 Synchronic changes of the (a) instant Doppler centroid and (b) water level in the tank (Rozenberg et al. 1973).

the shift of centre of gravity of the spectrum (spectrum centroid), and then the function spectrum $G_s(f)$ is found, one can obtain temporal spectrum of large wave elevations (this principle lies in the basis of microwave wavemeter with 1 km range measure capacity (Rozenberg 1980)) from the equation

$$G_\zeta(f) = \left(\frac{\lambda}{4\pi f}\right)^2 G_s(f) \tag{4.44}$$

So far we have talked about the signal Doppler spectrum for stationary radar. If radar is placed on the platform moving horizontally in the plane ZY with velocity V, then

$$G_0(\omega) \propto \langle\sigma_0^+\rangle \exp\left[-\frac{(\omega - \Omega_{\text{res}})^2}{8k^2\sigma_{\text{rad}}^2 + k^2V^2\delta_{y(2)}^2}\right] + \langle\sigma_0^-\rangle \exp\left[-\frac{(\omega + \Omega_{\text{res}})^2}{8k^2\sigma_{\text{rad}}^2 + k^2V^2\delta_{(2)}^2}\right] \tag{4.45}$$

where $\delta_{y(2)}$ is the two-way width of antenna pattern in the azimuthal plane at $1/e$ power level decrease. Evidently, at high velocity of the radar platform the spectrum width almost does not depend on the sea state and is determined by the product $V\delta_{y(2)}$. Nevertheless, the position of the spectral maximum can be determined with the accuracy of about 1 Hz even at satellite velocity (Li et al. 1982).

In conclusion, we point out that unlikely NRCS (i.e. backscattered signal intensity, in fact) Doppler spectrum parameters are independent of water temperature and salinity, which gives a certain advantage to remote probing.

– 5 –

Real aperture side-looking radar

As discussed in Section 3.2.2 and Chapter 4, we discussed the two-scale sea surface model or, more exactly, the microwave scattering composite theory and its verification with radar cross-section experimental data and scattered field Doppler spectrum. A verification study proved that the theory is completely valid in limits of its basic tenets validity. The backscatter composite theory application will be carefully examined, namely the development on its basis of the theory of radar imaging of the ocean surface.

This chapter studies the side-looking RAR imaging the surface, based on the spatio-temporal field of the returned signal intensity (or amplitude).

5.1 CORRELATION FUNCTION OF BACKSCATTERED SIGNAL INTENSITY

As discussed above, imaging radars operate in pulse mode. However, at the correctly chosen frequency of pulse repetition (rather high so as not to lose useful information in the reflected signal and at same time avoiding confusion due to the temporal overlapping of reflected signals) the emission can be seen as continuous during our analysis.

Let us return to Eqn. (4.6); assuming the antenna is in the far zone relative to the ground range radar resolution cell, i.e. fulfilling the condition

$$\frac{k(\Delta x)^2}{R} \ll 1 \tag{5.1}$$

we write down for the normalized intensity $I = a \cdot a^*$ of the reflected field,

$$
\begin{aligned}
I(\vec{r},t) = \frac{4k^4}{\pi} \int\int d\vec{r_1} d\vec{r_2}\ & \phi\left(\vec{r_1}-\vec{r}\right)\phi\left(\vec{r_2}-\vec{r}\right) \\
& \times m\left(\vec{r_1},t\right)m^*\left(\vec{r_2},t\right)\xi\left(\vec{r_1},t\right)\xi\left(\vec{r_2},t\right) \\
& \times \exp\left\{2ik\left[(x_1-x_2)\cos\psi_0 - \left[\zeta\left(\vec{r_1},t\right)-\zeta\left(\vec{r_2},t\right)\right]\right.\right. \\
& \left.\left. \times \sin\psi_0 + \frac{1}{2R}\left[(y_1-y)^2-(y_2-y)^2\right]\right]\right\}
\end{aligned}
\tag{5.2}
$$

Here $\phi\left(\vec{r}_{1,2} - \vec{r}\right)$ is the function significantly different from zero only within the resolution cell; to simplify the expression hereafter we shall denote the slant distance of the resolution cell central point by R instead of R_0, as used previously in Chapter 4.

Note that hereafter as well as sometimes earlier though (see the comment to expression (3.72)), speaking about the radar resolution cell, we shall use the "azimuthal size" term keeping in mind the two-way antenna pattern angle width in the azimuthal plane.

We set a spatio-temporal correlation function

$$B(\vec{\rho}, \tau) = \left\langle I(\vec{r}, t) I(\vec{r} + \vec{\rho}, t + \tau) \right\rangle \tag{5.3}$$

where $\vec{\rho} = \{\rho_x, V\tau\}$ and τ are spatial and temporal lags:

$$
\begin{aligned}
B(\vec{\rho}, \tau) = \left(\frac{4k^4}{\pi}\right)^2 \Bigg\langle & \iiiint d\vec{r}_1\, d\vec{r}_2 d\vec{r}_3\, d\vec{r}_4 \phi\left(\vec{r}_1 - \vec{r}\right) \phi\left(\vec{r}_2 - \vec{r}\right) \\
& \times \phi\left(\vec{r}_3 - (\vec{r} + \vec{\rho})\right) \phi\left(\vec{r}_4 - (\vec{r} + \vec{\rho})\right) \\
& \times m\left(\vec{r}_1, t\right) m^*\left(\vec{r}_2, t\right) m\left(\vec{r}_3, t + \tau\right) m^*\left(\vec{r}_4, t + \tau\right) \\
& \times \xi\left(\vec{r}_1, t\right) \xi\left(\vec{r}_2, t\right) \xi\left(\vec{r}_3, t + \tau\right) \xi\left(\vec{r}_4, t + \tau\right) \\
& \times \exp\Big\{ 2ik\Big[(x_1 - x_2 + x_3 - x_4)\cos\psi_0 \\
& \quad - \Big[\varsigma\left(\vec{r}_1, t\right) - \varsigma\left(\vec{r}_2, t\right) + \varsigma\left(\vec{r}_3, t + \tau\right) - \varsigma\left(\vec{r}_4, t + \tau\right)\Big]\sin\psi_0\Big]\Big\} \\
& \times \exp\Big\{\frac{ik}{R}\Big[(y_1 - y)^2 - (y_2 - y)^2\Big]\Big\} \\
& \times \exp\Big\{\frac{ik}{R_+}\Big[(y_3 - (y + V\tau))^2 - (y_4 - (y + V\tau))^2\Big]\Big\}\Bigg\rangle
\end{aligned}
\tag{5.4}
$$

where R_+ is the distance to the centre shifted by $\vec{\rho}$ of the resolution cell from the antenna location at the time of $t + \tau$.

Large-scale waves and "standard" ripples can evidently be taken as statistically independent. This means that the averaged part dependent on ξ becomes a separate multiplier, that is the fourth correlation moment. Assuming that the standard ripple distribution is a Gaussian one, we can write (see, e.g. Levin 1969)

$$
\begin{aligned}
\Big\langle \xi\left(\vec{r}_1, t\right) \xi\left(\vec{r}_2, t\right) \xi\left(\vec{r}_3, t + \tau\right) \xi\left(\vec{r}_4, t + \tau\right) \Big\rangle \\
= B_\xi\left(\vec{r}_1 - \vec{r}_2, 0\right) B_\xi\left(\vec{r}_3 - \vec{r}_4, 0\right) + B_\xi\left(\vec{r}_1 - \vec{r}_4, \tau\right) B_\xi\left(\vec{r}_2 - \vec{r}_3, \tau\right) \\
+ B_\xi\left(\vec{r}_1 - \vec{r}_3, \tau\right) B_\xi\left(\vec{r}_2 - \vec{r}_4, \tau\right)
\end{aligned}
\tag{5.5}
$$

where B_ξ is the ripple spatio-temporal correlation function. As a result, the correlation function of the backscattered field intensity falls into three summands:

$$B(\vec{\rho},\tau) = B_1(\vec{\rho},\tau) + B_2(\vec{\rho},\tau) + B_3(\vec{\rho},\tau) \tag{5.6}$$

We analyse the first summand and at the same time switch over to new variables:

$$\begin{aligned} \vec{r}_1 - \vec{r}_2 &= \vec{\rho}', & \vec{r}_3 - \vec{r}_4 &= \vec{\rho}'' \\ \vec{r}_1 + \vec{r}_2 &= 2\vec{r}', & \vec{r}_3 + \vec{r}_4 &= 2\vec{r}'' \end{aligned} \tag{5.7}$$

The previous variables are expressed in terms of new ones:

$$\begin{aligned} \vec{r}_1 &= \vec{r}' + \vec{\rho}'/2, & \vec{r}_3 &= \vec{r}'' + \vec{\rho}''/2 \\ \vec{r}_2 &= \vec{r}' - \vec{\rho}'/2, & \vec{r}_4 &= \vec{r}'' - \vec{\rho}''/2 \end{aligned} \tag{5.8}$$

The Jacobian of this transformation is zero.

After we expressed $B_1(\vec{\rho},\tau)$ through new variables bearing in mind that the functions ϕ and m in Eqn. (5.4) almost stay invariant on the ripple correlation spatial scale, we find

$$\begin{aligned} B_1(\vec{\rho},\tau) = \left(\frac{4k^4}{\pi}\right)^2 &\iint d\vec{r}'' \, d\vec{r}'' \, \Phi\left(\vec{r}'-\vec{r}\right)\Phi\left(\vec{r}''-(\vec{r}+\vec{\rho})\right) \\ &\times \left\langle \left|m(\vec{r}',t)\right|^2 \left|m(\vec{r}'',t+\tau)\right|^2 \iint d\vec{\rho}' \, d\vec{\rho}'' \, B_\xi(\vec{\rho}',0)B_\xi(\vec{\rho}'',0) \right. \\ &\times \exp\left\{2ik\left[\left(\cos\psi_0 - \frac{\partial\zeta}{\partial x}(\vec{r}',t)\sin\psi_0\right)\rho_x' + \left(\frac{y'-y}{R} - \frac{\partial\zeta}{\partial y}(\vec{r}',t)\sin\psi_0\right)\rho_y'\right]\right\} \\ &\times \exp\left\{2ik\left[\left(\cos\psi_0 - \frac{\partial\zeta}{\partial x}(\vec{r}',t+\tau)\sin\psi_0\right)\rho_x' \right.\right. \\ &\left.\left.+ \left(\frac{y''-(y+V\tau)}{R_+} - \frac{\partial\zeta}{\partial y}(\vec{r}'',t+\tau)\sin\psi_0\right)\rho_y''\right]\right\}\right\rangle \end{aligned} \tag{5.9}$$

Here we have used the symbol $\Phi = \phi^2$. It is easy to notice that the internal integrals break apart in Eqn. (5.9), and each part is nothing but the spatial spectrum of the ripples $\hat{W}_\xi$ at the local resonance spatial frequency defined by the local grazing angle and the azimuthal angle, at which the given point on the surface is seen from point of antenna location:

$$\int d\vec{\rho}' \, B_\xi(\vec{\rho}',0)\exp\left\{2ik\left[\left(\cos\psi_0 - \frac{\partial\zeta}{\partial x}(\vec{r}',t)\sin\psi_0\right)\rho_x' + \left(\frac{y'-y}{R} - \frac{\partial\zeta}{\partial y}(\vec{r}',t)\sin\psi_0\right)\rho_y'\right]\right\} = 4\pi^2 \hat{W}_\xi\left(\vec{\kappa}_{res}(\vec{r}',t)\right) \tag{5.10}$$

Hence:

$$B_1(\vec{\rho},\tau) = \langle \hat{\sigma}(\vec{r},t)\hat{\sigma}(\vec{r}+\vec{\rho},t+\tau)\rangle \tag{5.11}$$

$$\hat{\sigma}(\vec{r},t) = \int d\vec{r}' \, \Phi(\vec{r}'-\vec{r})\sigma_0(\vec{r}',t) \tag{5.12}$$

$$\sigma_0 = 16\pi k^4 |m|^2 \hat{W}_\xi\left(\vec{\kappa}_{\text{res}}\right) \tag{5.13}$$

Evidently, σ_0 is the mean backscatter cross-section of the quasi-flat facet on the surface $\zeta(\vec{r}, t)$ having a fixed slope covered by "standard" ripples; yet here contrary to Eqn. (3.10) the multiplier $|m|^2$ applies to both modulation types – geometrical and hydrodynamic.

Thus, $B_1(\vec{\rho}, \tau)$ describes slow (with a specific scale of the long wavelength and period) intensity fluctuations of the backscattered electromagnetic field pulse. In other words, B_1 is the spatio-temporal function of the useful signal correlation, i.e. the signal tracking the long wave. As for the sum $B_2 + B_3$, we shall so far define it as "the rest of the entire correlation function $B(\vec{\rho}, \tau)$ after its 'useful' member B_1 is subtracted".

Then we turn to $B_2(\vec{\rho}, \tau)$, the second summand on the right-hand side of Eqn. (5.6). After we have averaged over the ripple realizations and introduced correlation functions with arguments $\vec{r}_1 - \vec{r}_4$ and $\vec{r}_1 - \vec{r}_3$ according to Eqn. (5.5), the expression for B_2 becomes

$$B_2 = \langle \tilde{\sigma} \tilde{\sigma}^* \rangle \tag{5.14}$$

$$\begin{aligned}
\tilde{\sigma} = \frac{4k^4}{\pi} &\int\int d\vec{r}_1 \; d\vec{r}_4 \; \phi\left(\vec{r}_1 - \vec{r}\right)\phi\left(\vec{r}_4 - (\vec{r} + \vec{\rho})\right) \\
&\times \left\langle m\left(\vec{r}_1, t\right) m^*\left(\vec{r}_4, t + \tau\right) B_\xi\left(\vec{r}_1 - \vec{r}_4, \tau\right) \right. \\
&\times \exp\left\{2ik\left[(x_1 - x_4)\cos\psi_0 - \left[\zeta\left(\vec{r}_1, t\right) - \zeta\left(\vec{r}_4, t + \tau\right)\right]\sin\psi_0\right]\right\} \\
&\left. \times \exp\left[\frac{ik}{R}(y_1 - y)^2\right]\exp\left[\frac{ik}{R_+}(y_4 - (y + V\tau))^2\right]\right\rangle
\end{aligned} \tag{5.15}$$

In view of Eqn. (5.1), we write

$$R_+ = R + \rho_x \cos\psi_0 \tag{5.16}$$

Thus,

$$\frac{1}{R_+} \approx \frac{1}{R}\left(1 - \frac{\rho_x}{R}\cos\psi_0\right) \tag{5.17}$$

Integration in Eqn. (5.15) is over the overlapping part of resolution cells shifted against each other by $\vec{\rho}$; consequently, the linear dimensions of the integration area cannot exceed Δx along the X-axis and Δy along the Y-axis. Therefore if we replace R_+ by R, in Eqn. (5.15), the argument module in the last exponent does not change by more than

$$\Delta\varphi = \frac{k}{R}\frac{\Delta x}{R}(\Delta y)^2 = k\Delta x \delta_{y(2)}^2 \tag{5.18}$$

where $\delta_{y(2)}$ is the two-way angular width of the antenna pattern in the azimuthal plane. Setting $\Delta x = 10\,\text{m}$, $\delta_y = 5°$, for the electromagnetic wave $\lambda = 3\,\text{cm}$ we get $\Delta\varphi \approx \pi/40$. Hence, we can substitute R_+ by R, in Eqn. (5.15) without any significant inaccuracy; we will exactly follow this then.

We switch over to new variables:

$$\vec{r}_1 - \vec{r}_4 = \vec{\rho}\,', \quad \vec{r}_1 + \vec{r}_4 = 2\vec{r}\,' \tag{5.19}$$

Thereafter we have performed the same computations, this time for the relation

$$\int d\vec{\rho}\,' \; B_\xi(\vec{\rho}\,',\tau) \exp\left(-i\vec{\kappa}_{\text{res}} \vec{\rho}\,'\right) = 4\pi^2 \int d\omega \; \Psi_\xi\left(\vec{\kappa}_{\text{res}},\omega\right) \exp(-i\omega\tau) \tag{5.20}$$

where $\Psi_\xi\left(\vec{\kappa}_{\text{res}},\omega\right)$ is the ripple spatio-temporal spectrum, to obtain the result

$$B_2(\vec{\rho},\tau) = \langle \tilde{\sigma}(\vec{r},t;\vec{\rho},\tau)\tilde{\sigma}^*(\vec{r},t;\vec{\rho},\tau)\rangle \tag{5.21}$$

$$\tilde{\sigma} = 16\pi k^4 \int d\vec{r}\,' \; \phi(\vec{r}\,'-\vec{r})\phi\left(\vec{r}\,'-(\vec{r}+\vec{\rho})\right)\left|m(\vec{r}\,',t)\right|^2 \times \int d\omega \; \Psi_\xi\left(\vec{\kappa}_{\text{res}}(\vec{r}\,',t),\omega\right)$$
$$\times \exp\left\{-i\left[\omega - 2k\left[\frac{\partial \zeta}{\partial t}\left(\vec{r}\,',t\right)\sin\psi_0 + \frac{V}{R}\left(x'-x-\frac{V\tau}{2}\right)\right]\right]\right\} \tag{5.22}$$

This part of the entire correlation function $B(\vec{\rho},\tau)$, describing the signal fast fluctuations, is analysed in Section 5.3.

As for B_3, the third summand on the right-hand part of Eqn. (5.9), after we write an equation similar to Eqn. (5.9), we discover fast-oscillating multipliers in the integrand, so that the third summand turns out to be at least $(\lambda/\Delta x)^2$ times less than the first and second ones (Δx is the radar resolution cell size at the ground range).

5.2 SPATIAL SPECTRUM OF SIGNAL INTENSITY SLOW FLUCTUATIONS

We will take the correlation function of the signal intensity slow fluctuations which, in fact, image the large ocean waves:

$$B_1(\vec{\rho},\tau) = \left\langle \hat{\sigma}\left(\vec{r},t\right)\hat{\sigma}\left(\vec{r}+\vec{\rho},t+\tau\right)\right\rangle$$
$$= \int\int d\vec{r}\,' \, d\vec{r}\,'' \, \Phi\left(\vec{r}\,'-\vec{r}\right)\Phi\left(\vec{r}\,'' - \left(\vec{r}+\vec{\rho}\right)\right)\left\langle \sigma_0\left(\vec{r}\,',t\right)\sigma_0\left(\vec{r}\,'',t+\tau\right)\right\rangle \tag{5.23}$$

Switching over to variables $\vec{\rho}\,' = \vec{r}\,'' - \vec{r}\,'$ and $\vec{r}\,' = \vec{r}\,'$ we get

$$B_1(\vec{\rho},\tau) = \int d\vec{\rho}\,' \; b(\vec{\rho} - \vec{\rho}\,')B_0(\vec{\rho}\,',\tau) \tag{5.24}$$

$$b(\vec{\rho}) = \int d\vec{r}\,' \; \Phi(\vec{r}\,' - \vec{r}) \Phi\left(\vec{r}\,' - (\vec{r} + \vec{\rho})\right) \tag{5.25}$$

$$B_0(\vec{\rho}\,', \tau) = \langle \sigma_0(\vec{r}\,', t)\sigma_0(\vec{r}\,' + \vec{\rho}\,', t + \tau)\rangle \tag{5.26}$$

Using expressions

$$b(\vec{\rho}) = \int d\vec{\kappa}\, w(\vec{\kappa}) \exp(i\,\vec{\kappa}\,\vec{\rho}) \tag{5.27}$$

$$B_0(\vec{\rho}, \tau) = \int\int d\vec{\kappa}\, d\omega\, \Psi_0(\vec{\kappa}, \omega) \exp\left[i(\vec{\kappa}\vec{\rho} - \omega\tau)\right] \tag{5.28}$$

we introduce spectral functions $w(\vec{\kappa})$ and $\Psi_0(\vec{\kappa}, \omega)$. Substitution of Eqns (5.27) and (5.28) into Eqn. (5.24) we obtain

$$\Psi_1(\vec{\kappa}, \omega) = w(\vec{\kappa})\Psi_0(\vec{\kappa}, \omega) \tag{5.29}$$

where Ψ_1 stands for the spectrum of correlation function B_1, and multiplier

$$w(\vec{\kappa}) = 4\pi^2 w(\vec{\kappa}) = \int d\vec{\rho}\;\; b(\vec{\rho}) \exp(-i\,\vec{\kappa}\vec{\rho}) \tag{5.30}$$

is nothing but the spectral characteristics of linear filter, which corresponds to radar resolution cell.

Having used the MTF $T(\vec{\kappa})$ introduced in Chapter 3, the spectrum of the variable $\delta\sigma_0$ is written as

$$\Psi_0(\vec{\kappa}, \omega) = \langle \sigma_0\rangle^2 M(\vec{\kappa})\Psi_\zeta(\vec{\kappa}, \omega) \tag{5.31}$$

where $M(\vec{\kappa}) = \left|T(\vec{\kappa})\right|^2$, and Ψ_ζ is the spatio-temporal spectrum of large wave elevations. Hence, for the spectrum of NRCS slow fluctuations $\delta\sigma_0$, we deduce

$$\Psi_1(\vec{\kappa}, \omega) = \langle \sigma_0\rangle^2 w(\vec{\kappa})M(\vec{\kappa})\Psi_\xi(\vec{\kappa}, \omega) \tag{5.32}$$

Given

$$\Phi(\vec{r}\,' - \vec{r}) = \exp\left\{-4\left[\left(\frac{x'-x}{\Delta x}\right)^2 + \left(\frac{y'-y}{\Delta y}\right)^2\right]\right\} \tag{5.33}$$

then

$$b(\vec{\rho}) = \frac{\pi}{8}\Delta x\,\Delta y\;\exp\left\{-2\left[\left(\frac{\rho_x}{\Delta x}\right)^2 + \left(\frac{\rho_y}{\Delta y}\right)^2\right]\right\} \tag{5.34}$$

$$w(\vec{\kappa}) = \left(\frac{\pi}{4} \Delta x \ \Delta y\right)^2 \exp\left\{-\frac{1}{8}\left[(\kappa_x \Delta x)^2 + (\kappa_y \Delta y)^2\right]\right\} \qquad (5.35)$$

which shows that the resolution cell filters out relatively short microwaves $\Lambda \leq 2\Delta x, 2\Delta y$ long.

Obviously, the radar moving parallel to the -axis (see Figure 1.1) sees large wave spectrum as transformed as azimuthal spatial frequencies change to temporal:

$$\Psi_\zeta(\vec{\kappa}, \omega) = \hat{W}_\zeta^+(\vec{\kappa})\delta\left[\omega - (\Omega - \kappa_y V)\right] + \hat{W}_\zeta^-(-\vec{\kappa})\delta\left[\omega + (\Omega + \kappa_y V)\right] \qquad (5.36)$$

Remember (see Eqn. (3.15)) that $\hat{W}_\zeta^{\pm}$ are the halves of the spatial spectrum ensuring its central symmetry, as $\hat{W}_\zeta^+(\vec{\kappa}) = \hat{W}_\zeta^-(-\vec{\kappa})$. Each superscript $\pm$ stands for the sign of temporal frequency Ω determined by dispersion correlation with the wave number κ. Limiting ourselves to positive frequencies Ω, we get

$$\Psi_\zeta(\vec{\kappa}, \omega) = W_\zeta(\vec{\kappa})\delta\left[\omega - (\Omega - \kappa_y V)\right] \qquad (5.37)$$

where

$$W_\zeta(\vec{\kappa}) = 2\hat{W}_\zeta^+(\vec{\kappa}) \qquad (5.38)$$

is the spatial spectrum in which the direction of the wave vector $\vec{\kappa}$ corresponds to the wave propagation direction.

To switch over from Eqn. (5.32) to the spatio-temporal spectrum of the side-looking RAR signal, we should substitute Eqn. (5.32) into Eqn. (5.37) and then integrate Eqn. (5.37) over variable κ_y:

$$\Psi_{\text{RAR}}(\kappa_x, \omega) = w\left(\kappa_x, -\frac{\omega - \Omega}{V}\right) M\left(\kappa_x, -\frac{\omega - \Omega}{V}\right) W_\zeta\left(\kappa_x, -\frac{\omega - \Omega}{V}\right) \qquad (5.39)$$

Switching to Eqn. (5.37), we notice that at fairly high radar platform velocity the positive values of κ_y change to negative values of ω, and so, we assume $-\omega/V = \kappa_y'$. Remember that κ_y' is a temporal frequency unfolded spatially by the radar movement, i.e. κ_y' is not a real but a quasi-spatial frequency.

Therefore, the spectrum of the sea roughness image formed by side-looking RAR is represented by

$$W_{\text{RAR}}\left(\kappa_x, \kappa_y'\right) = w\left(\kappa_x, \kappa_y' + \frac{\Omega}{V}\right) M\left(\kappa_x, \kappa_y' + \frac{\Omega}{V}\right) W_\zeta\left(\kappa_x, \kappa_y' + \frac{\Omega}{V}\right) \qquad (5.40)$$

Formula (5.40) is explained by the principle of the monochromatic swell imaging by the RAR, i.e. let $W_\zeta \propto \delta(\vec{\kappa} - \vec{\kappa}_0)$; then Eqn. (5.40) yields

$$W_{\text{RAR}}\left(\kappa_x, \kappa'_y\right) \propto w(\vec{\kappa}_0)M(\vec{\kappa}_0)\delta\left(\kappa_x - \kappa_{0x}, \kappa'_y - \left(\kappa_{0y} - \frac{\Omega_0}{V}\right)\right) \qquad (5.41)$$

Consequently, the scan-distorted azimuthal wave number is

$$\kappa'_{0y} = \kappa_{0y} - \frac{\Omega_0}{V} \qquad (5.42)$$

whereas the scanning does not distort the ground range wave number. Therefore, we can write

$$\kappa'_0 \cos \phi'_0 = \kappa_0 \left(\cos \phi_0 - \frac{V_{\text{ph}}}{V} \right) \qquad (5.43)$$

$$\kappa'_0 \sin \phi'_0 = \kappa_0 \sin \phi_0 \qquad (5.44)$$

where κ'_0 is the apparent swell wave number, ϕ_0 is the angle between the real wave vector $\vec{\kappa}_0$ and the Y-axis, ϕ'_0 is the same for the apparent wave vector $\vec{\kappa}'_0$, and $V_{\text{ph}} = \Omega_0/\kappa_0$ is the swell phase velocity. Both ϕ_0 and ϕ'_0 are taken positive clockwise from the Y-axis direction. It can be seen easily from Eqns (5.43) and (5.44) that

$$\phi'_0 = \tan^{-1}\left(\frac{\sin \phi_0}{\cos \phi_0 - V_{\text{ph}}/V}\right) \qquad (5.45)$$

Having defined a non-dimensional distortion parameter ε (see Rufenach et al. 1991)

$$\varepsilon = 2\frac{V_{\text{ph}}}{V} \cos \phi_0 \qquad (5.46)$$

one can find the distorted wave number

$$\kappa'_0 = \kappa_0 (1 - \varepsilon)^{1/2} \qquad (5.47)$$

and the distorted swell wavelength

$$\Lambda'_0 = \frac{\Lambda_0}{(1 - \varepsilon)^{1/2}} \qquad (5.48)$$

The scanning distortions described by Eqns (5.45), (5.47) and (5.48) are dependent on the ratio of the ocean wave phase velocity to the platform velocity. This distortion as a function of azimuth angle α is given in Figure 5.1 (quoted from Rufenach et al. 1991). The calculations have been carried out for the radar platform velocity $V = 100\,\text{m s}^{-1}$. The

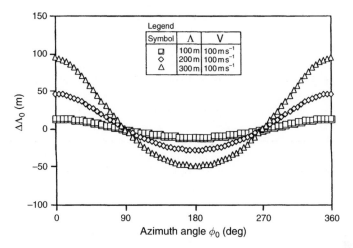

Figure 5.1 Differential wavelength scanning distortion as a function of aircraft heading relative to the direction of ocean wave travel, azimuthal angle α. The results are parameterized in typical gravity wavelengths $\Lambda = 100$ m (1), 200 m (2) and 300 m (3) and the aircraft velocity $V = 100$ m s^{-1} (Rufenach et al. 1991).

peak of wavelength distortion $\Delta\Lambda_0 = \Lambda_0' - \Lambda_0$ is substantial for the wave travelling in or near the azimuth direction ($\phi_0 \approx 0°$) with distortion up to 50 or 100 m for slow-flying aircraft and wavelength greater than 250 m, whereas for waves traveling in or near the range direction ($\phi_0 \approx 90°$) the wavelength distortion is minimum. The distorted wavelength is longer than the actual wavelength when the wave and the radar platform move in the same direction and shorter when they are in opposite directions. The maximum direction (angle) distortion is about 10° as illustrated in Figure 5.2 (quoted from Rufenach

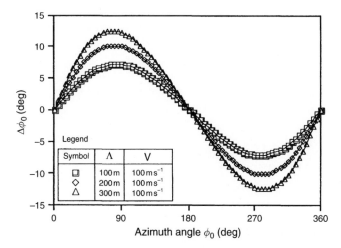

Figure 5.2 Differential directional scanning distortion as a function of azimuthal angle α. The results are parameterized in the typical gravity wavelengths $\Lambda = 100$ m (1), 200 m (2) and 300 m (3) and the aircraft velocity $V = 100$ m s^{-1} (Rufenach et al. 1991).

et al. 1991). The maximum clockwise rotation corresponds to waves travelling in the range direction, and the maximum counterclockwise rotation corresponds to waves travelling opposite the range direction. The ocean wave rotation is minimal when the ocean waves are traveling in the same direction as the radar platform moves in.

Evidently, the distortions disappear at the radar platform high velocity, and ignoring the MTF influence and the resolution cell restrictions the apparent roughness is almost the same as the actual one.

5.3 FAST SIGNAL FLUCTUATIONS (SPECKLE NOISE)

Let us consider the part of the signal described by the summand B_2 of the full correlation function (5.6). We write the spatio-temporal spectrum of the $\Psi_\xi(\vec{\kappa}_{\text{res}}, \omega)$ ripples (the value is included in Eqn. (5.21)) as

$$
\Psi_\xi\left(\vec{\kappa}_{\text{res}}, \omega\right) = \hat{W}_\xi^+\left(\vec{\kappa}_{\text{res}}\right)\delta\left(\omega - \left(\Omega_{\text{res}} + \vec{\kappa}_{\text{res}}\,\vec{v}_{\text{orb}}\right)\right)
$$
$$
+ \hat{W}_\xi^-\left(\vec{\kappa}_{\text{res}}\right)\delta\left(\omega + \left(\Omega_{\text{res}} + \vec{\kappa}_{\text{res}}\,\vec{v}_{\text{orb}}\right)\right)
\tag{5.49}
$$

Introducing only positive temporal frequencies, we can rewrite Eqn. (5.49) as

$$
\Psi_\xi\left(\vec{\kappa}_{\text{res}}, \omega\right) = \hat{W}_\xi^+\left(\vec{\kappa}_{\text{res}}\right)\delta\left(\omega - \left(\Omega_{\text{res}} + \vec{\kappa}_{\text{res}}\,\vec{v}_{\text{orb}}\right)\right)
$$
$$
+ \hat{W}_\xi^+\left(-\vec{\kappa}_{\text{res}}\right)\delta\left(\omega - \left(\Omega_{\text{res}} - \vec{\kappa}_{\text{res}}\,\vec{v}_{\text{orb}}\right)\right)
\tag{5.50}
$$

Note that Eqn. (5.49), unlike Eqn. (5.36), describes two counter-propagating ripple waves, namely, approaching to the radar and receding from it. Obviously, in terms of the scatter both waves are quite equivalent in the sense that the intensity of the electromagnetic scatter of each ripple wave is defined only by this wave intensity, rather than by its propagation direction. Generally, the resonance ripple temporal frequency is evidently different from Ω_{res}, as ripples are part of the orbital movement in the large-scale roughness field and therefore are characterized by the $\vec{v}_{\text{orb}}$ orbital velocity besides the proper phase one.

Remember that the two counter-propagating resonance components were dealt with in Chapter 3 (see Figure 3.6) when speaking of the HF diapason; the difference in the intensity of the two waves was about 20 dB. Again, we are assuming the ripple spectrum components as insignificantly small whose propagating direction has a negative projection onto the wind direction, consequently,

$$
\Psi_\xi\left(\vec{\kappa}_{\text{res}}, \omega\right) = \hat{W}_\xi^+\left(\vec{\kappa}_{\text{res}}\right)\delta\left(\omega - \left(\Omega_{\text{res}} + \vec{\kappa}_{\text{res}}\,\vec{v}_{\text{orb}}\right)\right)
\tag{5.51}
$$

Taking into account that $\vec{\kappa}_{\text{res}} = \{-2k\cos\psi_0, 0\}$, one can see that Eqn. (5.51) has been written by assuming $\vec{k}\,\vec{U} < 0$, where $\vec{U}$ is the wind speed vector. The allowances we made for Eqn. (5.51) are not crucial, and were introduced only to achieve simplification.

Inserting Eqn. (5.51) into Eqn. (5.22), we integrate the ultimate expression over ω with allowance for the following relation:

$$\vec{\kappa}_{\text{res}}\vec{v}_{\text{orb}} + 2k\sin\psi_0\frac{\partial\zeta}{\partial t} = -2\,\vec{k}\,\vec{v}_{\text{orb}} = 2kv_{\text{rad}} \tag{5.52}$$

where v_{rad} is the radial component of the orbital velocity, and obtain

$$B_2(\vec{\rho},\tau) = \int\int \mathrm{d}\vec{r}'\,\mathrm{d}\vec{r}''\,\phi(\vec{r}'-\vec{r})\phi(\vec{r}''-\vec{r})\phi\left(\vec{r}'-(\vec{r}+\vec{\rho})\right)\phi\left(\vec{r}''-(\vec{r}+\vec{\rho})\right)$$
$$\times\exp\left\{2ik\frac{V}{R}(y'-y'')\tau\right\}\left\langle\sigma_0(\vec{r}')\sigma_0(\vec{r}'')\exp\left\{2ik\left[v_{\text{rad}}(\vec{r}')-v_{\text{rad}}(\vec{r}'')\right]\tau\right\}\right\rangle \tag{5.53}$$

We further average expression (5.53) assuming

$$\left\langle\sigma_0(\vec{r}')\sigma_0(\vec{r}'')\exp\left\{2ik\left[v_{\text{rad}}(\vec{r}')-v_{\text{rad}}(\vec{r}'')\right]\tau\right\}\right\rangle$$
$$\approx\langle\sigma_0(\vec{r}')\sigma_0(\vec{r}'')\rangle\left\langle\exp\left\{2ik\left[v_{\text{rad}}(\vec{r}')-v_{\text{rad}}(\vec{r}'')\right]\tau\right\}\right\rangle \tag{5.54}$$

This assumption does not seem to affect further computation crucially. Moreover, as we shall see later, the results from the expression turned out quite physically transparent and in good agreement with experimental data.

Since the random ocean surface elevations $\zeta(\vec{r},t)$ have the Gaussian distribution, then both the orbital velocity and its radial component are also characterized by Gaussian distribution. Thus

$$\left\langle\exp\left\{2ik\left[v_{\text{rad}}(\vec{r}')-v_{\text{rad}}(\vec{r}'')\right]\tau\right\}\right\rangle = \exp\left\{-4k^2\sigma_{\text{rad}}^2\left[1-b_{\text{rad}}(\vec{r}'-\vec{r}'')\right]\tau^2\right\} \tag{5.55}$$

where σ_{rad}^2 and b_{rad} are the variance and the coefficient of the correlation of the orbital velocity radial component.

Switching over to the $\vec{r}'$ and $\vec{\rho}' = \vec{r}' - \vec{r}''$ variables, we obtain

$$B_2(\vec{\rho},\tau) = \int \mathrm{d}\vec{\rho}'\,b(\vec{\rho},\vec{\rho}')B_0(\vec{\rho}')\exp\left(2ik\frac{V}{R}\rho_y'\tau\right)$$
$$\times\exp\left\{-4k^2\sigma_{\text{rad}}^2\left[1-b_{\text{rad}}(\vec{\rho}')\right]\tau^2\right\} \tag{5.56}$$

where

$$b(\vec{\rho},\vec{\rho}') = \int \mathrm{d}\vec{r}'\,\phi(\vec{r}'-\vec{r})\phi\left(\vec{r}'-(\vec{r}+\vec{\rho})\right)$$
$$\times\phi\left(\vec{r}'-(\vec{r}+\vec{\rho}')\right)\phi\left(\vec{r}'-(\vec{r}+\vec{\rho}+\vec{\rho}')\right) \tag{5.57}$$

Approximation (5.33) for $b(\vec{\rho}, \vec{\rho}\,')$ yields

$$b(\vec{\rho}, \vec{\rho}\,') = \frac{\pi}{8} \Delta x \, \Delta y \ \exp\left\{-2\left[\frac{\rho_x^2 + \rho_x'^2}{(\Delta x)^2} + \frac{\rho_y^2 + \rho_y'^2}{(\Delta y)^2}\right]\right\} \tag{5.58}$$

Further analysis will be based on the following assumptions. First, we will assume the radar resolution cell size as small compared to the surface wave characteristic length. Secondly, we will suppose that the probing is effected along or against the general wave direction. Thus, we take the multiplier $B_0(\vec{\rho}\,') \approx B_0(0) = \langle \sigma_0^2 \rangle$ outside the integral sign and write

$$B_2(\vec{\rho}, \tau) = \langle \sigma_0^2 \rangle \int d\vec{\rho}\,' \ b(\vec{\rho}, \vec{\rho}\,') \exp\left(2ik\frac{V}{R}\rho_y'\tau\right)$$
$$\times \exp\left\{-4k^2\sigma_{\text{rad}}^2\left[1 - b_{\text{rad}}(\rho_x', 0)\right]\tau^2\right\} \tag{5.59}$$

We take $B_0(\vec{\rho})$ as almost invariable for the resolution cell size, but the conjecture is generally not quite true for the correlation coefficient b_{rad}. In fact, the surface elevations spectrum and the orbital velocities spectrum are generally not the same and the corresponding wave characteristic lengths may not coincide. Their coincidence might take place solely at the quasi-monochromatic swell, while in fully developed windsea a characteristic surface wavelength may be more than three times that of the characteristic wave in the orbital velocities spectrum (for a more detailed analysis see Section 4.4). Therefore, taking the first derivative from $R_k(\vec{\rho})$ with $\vec{\rho} = 0$ as equal to zero, we get

$$b_{\text{rad}}(\rho_x, 0) \approx 1 - \frac{1}{2}\left|b_{\text{rad}}''(0)\right|\rho_x^2 \tag{5.60}$$

where b_{rad}'' is the second-order derivative with respect to ρ_x at $\vec{\rho} = 0$. Thus,

$$B_2(\vec{\rho}, \tau) = \langle \sigma_0^2 \rangle \int d\vec{\rho}\,' \ b(\vec{\rho}, \vec{\rho}\,') \exp\left(2ik\frac{V}{R}\rho_y'\tau\right)$$
$$\times \exp\left[-2k^2\sigma_{\text{rad}}^2\left|b_{\text{rad}}''(0)\right|\rho_x'^2\tau^2\right] \tag{5.61}$$

We insert expression (5.58) into (5.61), replace ρ_y by $V\tau$, and find integral over $\vec{\rho}\,'$ and obtain

$$B_2(\rho_x, \tau) = \frac{\pi}{16}\langle \sigma_0^2 \rangle (\Delta x \Delta y)^2 \exp\left[-\frac{2\rho_x^2}{(\Delta x)^2}\right]$$
$$\times \frac{\exp\left\{-[2V^2/(\Delta y)^2 + 0.5k^2(V/R)^2(\Delta y)^2]\tau^2\right\}}{\left[1 + k^2\sigma_{\text{rad}}^2\left|b_{\text{rad}}''(0)\right|(\Delta x)^2\tau^2\right]^{1/2}} \tag{5.62}$$

Let us compare the summands in the second-order exponent index:

$$\frac{2}{(\Delta y)^2} : \frac{1}{2} k^2 \left(\frac{\Delta y}{R} \right)^2 = \frac{1}{\pi^2} \left(\frac{\sqrt{\lambda R}}{\Delta y} \right)^4 << 1 \tag{5.63}$$

The latter inequation works due to the resolution cell location in the antenna far zone. Therefore

$$B_2(\rho_x, \tau) = \frac{\pi}{16} (\Delta x \Delta y)^2 \langle \sigma_0^2 \rangle \exp\left[-\frac{2\rho_x^2}{(\Delta x)^2} \right]$$

$$\times \frac{\exp[-0.5 k^2 V^2 \delta_{y(2)}^2 \tau^2]}{[1 + k^2 \sigma_{\text{rad}}^2 |b_{\text{rad}}''(0)| (\Delta x)^2 \tau^2]^{1/2}} \tag{5.64}$$

where $\delta_{y(2)} = \Delta y / R$ is the two-way angular width of the antenna pattern in the azimuthal plane at the $1/e$ power level.

Let us turn to the relation

$$\left| b_{\text{rad}}''(0) \right| = \frac{1}{\sigma_{\text{orb}}^2} \int \int d\kappa_x\, d\kappa_y \kappa_x^2 \hat{W}_{\text{v}}(\kappa_x, \kappa_y) \tag{5.65}$$

where $\hat{W}_{\text{v}}$ is the spatial spectrum of orbital velocities. Note that in our case when the observation is carried out perpendicularly to the wavefront the variances of the orbital velocity and its radial component almost coincide. The right-hand side of Eqn. (5.65) can be regarded as the average square of the wave number defining the wave characteristic length in the orbital velocities spectrum in the microwave incidence plane:

$$\left| b_{\text{rad}}''(0) \right| = \langle \kappa_x^2 \rangle = \left(\frac{2\pi}{\Lambda_{\text{v}}} \right)^2 \tag{5.66}$$

We introduce the following parameter:

$$\gamma_x = \frac{\Delta x}{\Lambda_0} \approx \frac{\Delta x}{3\Lambda_{\text{v}}} \tag{5.67}$$

where Λ_0 is the surface wave characteristic length at fully developed wind sea, and with regard to Eqn. (5.66), we have

$$B_2(\rho_x, \tau) = \frac{\pi}{16} (\Delta x \Delta y)^2 \langle \sigma_0^2 \rangle \exp\left[-\frac{2\rho_x^2}{(\Delta x)^2} \right] \frac{\exp\left[-0.5 k^2 V^2 \delta_{y(2)}^2 \tau^2 \right]}{\left[1 + k^2 (6\pi \gamma_x \sigma_{\text{rad}})^2 \tau^2 \right]^{1/2}} \tag{5.68}$$

It is useful to further compare $V\delta_{y(2)}$ and $6\pi\gamma_x\sigma_{\text{rad}}$. We assume $V = 200$ m s^{-1} and $\delta_{y(2)} = 0.5°$, i.e. $V\delta_{y(2)} = 1.7$ m s^{-1}. Let $\Lambda_0 = 100$ and $\Delta x = 5$ m; then at $\sigma_{\text{rad}} = 0.7$ m s^{-1} we obtain $6\pi\gamma_x\sigma_{\text{rad}} = 0.66$ m s^{-1}, i.e. $V\delta_{y(2)} \approx 2.6(6\pi\gamma_x\sigma_{\text{rad}})$. Thus with the exponent index varying from 0 to 2 the $0.5k^2(6\pi\gamma_x\sigma_{rad})^2\tau^2$ value changes from 0 to 0.3. Then we use the proximity of the $(1 + 2x^2)^{-1/2}$ and $\exp(-x^2)$ functions within the $0 \leq x \leq 0.3$ interval to our advantage and write Eqn. (5.68) as

$$
B_2(\rho_x, \tau) = \frac{\pi}{16}(\Delta x \Delta y)^2 \langle \sigma_0^2 \rangle \exp\left[-\frac{2}{(\Delta x)^2}\rho_x^2\right]
$$
$$
\times \exp\left\{-\frac{1}{2}k^2\left[(V\delta_{y(2)})^2 + (6\pi\gamma_x\sigma_{rad})^2\right]\tau^2\right\}
$$

(5.69)

Switching over from Eqn. (5.69) to the spatio-temporal spectrum denoted by Ψ^s_{RAR} we get

$$
\Psi^s_{RAR}(\kappa_x, \omega) = \frac{1}{64}\langle \sigma_0^2 \rangle \frac{(\Delta x)^3(\Delta y)^2}{\left[(V\delta_{y(2)})^2 + (6\pi\gamma_x\sigma_{rad})^2\right]^{1/2}}
$$
$$
\times \exp\left[-\frac{1}{8}(\kappa_x\Delta x)^2\right] \exp\left\{-\frac{\omega^2}{2k^2\left[(V\delta_{y(2)})^2 + (6\pi\gamma_x\sigma_{rad})^2\right]}\right\}
$$

(5.70)

Note that the Gaussian form of the spectrum temporal part was arrived at through the comparative values calculated above, which are strictly speaking of no fundamental nature. They have just allowed us to achieve the ultimate result more conveniently for the physical treatment shape.

The result is essentially as follows. The temporal fluctuations described by the summand B_2 of Eqn. (5.6) for the correlation function of the RAR signal intensity are caused by the beats resulting from the summation of the fields backscattered from resolution of cell different sections and characterized by different Doppler frequency shifts. The Doppler shift is explained by the following two factors. The first one is the shift caused by radar displacement and proportional to the incidence wave vector projection on radar direction. The corresponding Doppler frequencies range is evidently the broader the wider the antenna pattern in the azimuthal plane. The second factor is the orbital velocities of the Doppler-shifted scattering ripples. The Doppler bandwidth is proportional here to the scope of radial values of the orbital velocity within radar resolution cell.

As the spectrum width Δf_a of the amplitude temporal fluctuations is $\sqrt{2}$ times less than the corresponding intensity width, we have the following expression:

$$
\Delta f_a = \frac{1}{\lambda}\left[(V\delta_{y(2)})^2 + (6\pi\gamma_x\sigma_{rad})^2\right]^{1/2}
$$

(5.71)

For microwave probing with all the calculated data for Eqn. (5.71) values applicable, we get $\Delta f_a \approx 60$ Hz for airborne radar and $\Delta f_a \approx 20$ Hz for stationary radar set. Clearly, with the less focused antenna pattern and more extended in range direction resolution cell, the amplitude fluctuation spectrum will be broader. In particular, for $\delta_{y(2)} = 1°$ (one-way antenna pattern width $\delta_y \approx 1.4°$) and $\Delta x = 10$ m we obtain, respectively, $\Delta f_a \approx 120$ Hz and $\Delta f_a \approx 40$ Hz.

The ocean RAR-based spatial image features speckles formed by the signal temporal fluctuations; the respective speckle structure is the manifestation of the so-called speckle noise.

Expression (5.69) brings forth the conclusion that the characteristic X-component of the speckle is the resolution cell size Δx along the ground range. As for the speckle azimuth size (Y-component) under the conditions of fast probing, i.e. for airborne rather than ship-borne radar, it equals $\lambda/\delta_y \approx L$, which is the along way antenna size.

The speckle structure spatial spectrum can be obtained by just setting $\omega = \kappa_y' V$ in Eqn. (5.70) and thus transform the $\Psi^s_{RAR}(\kappa_x, \omega)$ spatio-temporal spectrum to the $\Psi^s_{RAR}(\kappa_x, \kappa_y')$ spatial spectrum.

Remember that the speckle noise by no means has an additive character; it is a multiplicative noise as backscattered fields with different Doppler frequencies arrive at the antenna having already undergone "slow" amplitude modulation, caused by large-scale waves. Due to the beats the antenna output fields are just as slowly modulated by amplitude. The additive character of the image full correlation function (5.6) should not mislead us here. Remember that we have supra-defined $B_2 + B_3$ as "the rest of the entire correlation function after its 'useful' member B_1 is subtracted". Moreover, due to the insignificantly small value of B_3, the definition also holds true for B_2. The analysis performed shows that the B_2 function has a certain connection to the multiplicative speckle noise. Yet what is the connection?

Let us present the full signal σ as $\sigma = \sigma_0 n$, where σ_0 and n are the useful signal and speckle noise, respectively. We will define a correlation function B of the σ value and take into account that σ_0 and n are statistically independent and besides $\langle n \rangle = 1$. Thus we get

$$B = B_1 + C_n B_1 \qquad (5.72)$$

where C_n is the covariance (i.e. correlation function of the fluctuations) of speckle noise. Therefore, $B_2 = C_n B_1$ and, consequently, the speckle underlayer spectrum beneath the large-scale roughness image is the convolution of the respective image and speckle-noise spectra. Since the characteristic spatio-temporal scale of the speckle noise is small as compared to that of the useful signal,

$$B = B_1 + \langle \sigma_0^2 \rangle C_n \qquad (5.73)$$

This means that the underlayer spectrum coincides with the speckle-noise spectrum yet is characterized by the coefficient equal to the average intensity of the useful signal, hence the additivity concept is out of the question.

5.4 INHOMOGENEOUS ROUGHNESS IMAGING

The ocean roughness has hitherto been viewed as homogeneous with the invariable statistic characteristics in the observed parts of the ocean surface. This generally is not the rule in real life (see Section 2.4). In fact, the ocean images made from space display inhomogeneities with wide-ranging area scales (from dozens and hundreds of metres in area to lots of kilometres) on the ocean surface. These inhomogeneities stem from various surface, subsurface and atmospheric processes.

Such phenomena as internal waves rise, currents, surface films and others modulate the roughness spectrum, which in turn changes the ocean reflecting capacity. In case of low-contrast formations caused among other factors by the low-amplitude internal waves, the main impact is on the small-scale gravity–capillary part of spectrum (ripples), which govern the centimetre and decimetre scatter.

The ripple modification by the internal waves has been analysed in a number of works (e.g. see Basovich et al. (1985) and references therein). This modification is measured by the hydrodynamic contrast K defined by the following relation:

$$W_\xi^{(1)}(\vec{\kappa};\vec{r},t) = \left[\langle K \rangle + \delta K(\vec{r},t)\right]W_\xi(\vec{\kappa};\vec{r},t) \qquad (5.74)$$

where W_ξ and $W_\xi^{(1)}$ are the ripple spectra disturbed and undisturbed, respectively, by the internal waves, and $\langle K \rangle$ and δK are the constant and variable parts of the hydrodynamic contrast. The slow spatio-temporal dependency of the W_ξ undisturbed spectrum originates from the large-scale waves effect on the ripples; the respective incidental fluctuations of the W_ξ spectrum, which are larger in scale against the wave period and length, shall be further held statistically independent to the still higher fluctuations of the contrast K brought forth by the internal waves.

As the microwave resonance scatter theory yields (see Chapter 3), the ripple spectrum disturbances bring about a modulating multiplier into the σ_0 cross section equation

$$\sigma_0^{(1)}(\vec{r},t) = \left[\langle K \rangle + \delta K(\vec{r},t)\right]\sigma_0(\vec{r},t) \qquad (5.75)$$

We take the spatial scale of the external disturbances correlation as considerably exceeding the linear size of the radar resolution cell. Thus with regard to the K contrast statistic independence, on the one hand, and the homogeneous roughness characteristics on the other, we have the following expression:

$$B_1^{(1)} = B_K B_1 \qquad (5.76)$$

$$B_2^{(1)} = \left[\langle K \rangle^2 + \left\langle (\delta K)^2 \right\rangle\right]B_2 \qquad (5.77)$$

where B_1 and B_2 are the correlation functions of the image proper and the speckle underlayer, and B_K is the contrast correlation function; the upper index applies to the disturbed roughness case.

For the fluctuation correlation functions (i.e. covariance) of the image proper and the speckle noise, Eqns (5.76) and (5.77) yield

$$b_1^{(1)} = \langle K \rangle^2 b_1 + \langle I_1 \rangle^2\, b_K + b_1 b_K \qquad (5.78)$$

$$b_2^{(1)} = \left(\langle K \rangle^2 + \sigma_K^2\right)b_2 \qquad (5.79)$$

where b_K is the contrast fluctuations correlation function, $\sigma_K^2 = \langle (\delta K)^2 \rangle$.

Switching over from Eqns (5.78) and (5.79) to the spatial spectra, we have

$$\hat{W}_{\mathrm{RAR}}^{(1)}(\vec{\kappa}) = \langle K \rangle^2 \hat{W}_{\mathrm{RAR}}(\vec{\kappa}) + \langle I_1 \rangle^2 \hat{W}_K(\vec{\kappa}) + \hat{W}_{\mathrm{RAR}} \otimes \hat{W}_K \qquad (5.80)$$

$$\hat{W}_{\mathrm{RAR}}^{\mathrm{s}(1)} = \left(\langle K \rangle^2 + \sigma_K^2 \right) \hat{W}_{\mathrm{RAR}}^{\mathrm{s}} \qquad (5.81)$$

where the superscript "s" denotes speckle-noise spectra.

As Eqn. (5.72) shows, with small σ_K^2 and large-scale external disturbances the speckle-noise spectrum is practically invariable. As for the two spectra convolution denoted by the $\otimes$ sign, it is explained by the non-additive character of the signal fluctuations, caused by the homogeneous large-scale roughness and external ripple disturbances. The physics of the convolution item is clarified by the supposition that the large-scale roughness is in fact monochromatic swell, i.e.

$$\hat{W}_\zeta(\vec{\kappa}) \propto \sum_{\mp} \delta \left(\vec{\kappa} \mp \vec{\kappa}_0 \right) \qquad (5.82)$$

Then

$$\hat{W}_{\mathrm{PPA}} \otimes \hat{W}_K \propto \sum_{\mp} \hat{W}_K \left(\vec{\kappa} \mp \vec{\kappa}'_0 \right) \qquad (5.83)$$

where $\vec{\kappa}'_0$ is the apparent wave number of the swell (see Eqn. (5.42)). Thus, the radar image of the ocean surface includes the foreign inhomogeneities spectrum shifted to the "carrier" swell spatial frequency as estimated by finite speed radar. (The "carrier frequency" term has been used here to highlight the parallel to amplitude modulation well-known in radio engineering. In this case, the modulating signal is the spatial modulation of the ripple intensity induced by external processes, and the "carrier frequency" is the apparent spatial frequency of the swell.) Therefore, the inhomogeneities are manifested not only in the spectral area corresponding to their size, but also in higher spatial frequency area (Kanevsky 1982, 1985).

Figures 5.3 and 5.4 exemplify the situation by image spectra of the adjoining surface areas; the data have been procured with ship-borne 3-cm radar during the expedition of the research group of Russian Institute of Applied Physics in the Atlantic Ocean (1992). In Figure 5.3 the ocean swell is betrayed in the symmetrical image spectrum as two sharp peaks. Figure 5.4 has an essentially different spectrum of the same swell due to the rather moderate internal waves. The peaks are broader since the convolution of the swell spectra and internal waves overlay the image spectrum proper, as shown in Figure 5.3. Hence, the apparent swell spectrum mutation in the presence of internal waves is nothing but an artefact, and by no means is a result of interaction between the swell and internal waves.

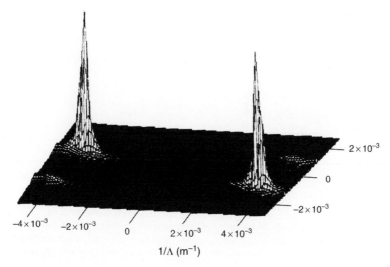

Figure 5.3 The spectrum of the "pure" swell image.

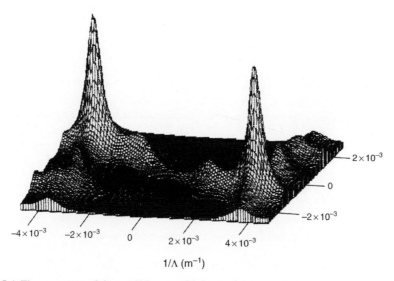

Figure 5.4 The spectrum of the swell image with internal waves present.

The example featured illustrates the necessity to adopt a careful approach in the interpretation of the results of the large-scale roughness radar imaging. Indeed, after the high frequency speckle-noise filtration the spectra of the roughness and its image may turn to be in the integral equation interrelation (5.80) rather then in the MTF one (see Chapter 3).

– 6 –

Synthetic aperture radar

SAR images the scattering surface from large distances, particularly, from space when the target resolution cannot be achieved by enlarging the physical aperture of the radar antenna. Here we are speaking about resolution over azimuthal coordinate, as the range resolution just as with RAR can be reached by deploying either rather short pulses, or chirp pulses with further compression (see Chapter 1). Basic concepts of SAR Earth surface imaging are covered in Chapter 1; this chapter is devoted to the specifics of ocean surface random movement.

6.1 PRELIMINARY ESTIMATES

We have mentioned above that SAR as an ocean surface probing tool is not in any way equivalent to incoherent side-looking radar having hypothetically super high resolution. To obtain preliminary estimates of roughness influence on SAR imaging of the ocean surface we consider expression (1.4):

$$
a_{\mathrm{SAR}}(t) = \frac{1}{\Delta t} \int\limits_{t-\Delta t/2}^{t+\Delta t/2} \mathrm{d}t' a(t') \exp\left[-\mathrm{i}\frac{k}{R} V^2 (t'-t)^2 \right] \tag{6.1}
$$

where $a(t)$ and $a_{\mathrm{SAR}}(t)$ are the complex amplitudes of backscattered field and of SAR signal, respectively; and Δt is the integration time (probing geometry is shown in Figure 1.1).

Having still the notion of SAR as a linear system it is easy to prove that the roughness-associated phenomena are the shift and expanding of the resolution cell.

In expression (1.8) for the complex amplitude of backscattered field

$$
a(t') \propto \mathrm{e}^{2\mathrm{i}kR} \int\limits_{\Delta y} \mathrm{d}y' p(y') \exp\left[\mathrm{i}\frac{k}{R} (y'-Vt')^2 \right] \tag{6.2}
$$

we assume $p(y') \propto \delta(y'-y_0) \exp(-2ikv_{\mathrm{rad}}t')$, where v_{rad} is the radial (i.e., along the radar-look direction) component (against the probing direction) of point target (scatterer) velocity; the component is held positive if directed to radar and negative if directed from it. Then, if $v_{\mathrm{rad}} = \mathrm{const.}$, then we get for SAR signal intensity:

$$I_{\mathrm{SAR}} = |a_{\mathrm{SAR}}(t)|^2 \propto \left(\frac{\sin u}{u} \right)^2 \tag{6.3}$$

$$u = \frac{kV\Delta t}{R} \left(y_0 - Vt + \frac{R}{V} v_{\mathrm{rad}} \right) \qquad .$$

i.e. the point scatterer image is shifted against y_0 by the range Rv_{rad}/V in either direction depending on the sign of v_{rad}. Thus, the image shifts along SAR carrier flight direction as the scatterer approaches radar and against as the scatterer moves away from radar.

We separate from Eqn. (6.2) the expressions for the Doppler frequency of the signal backscattered from the point scatterer:

$$\omega_{\mathrm{D}} = \begin{cases} -2\dfrac{k}{R}V(y_0 - Vt), & v_{\mathrm{rad}} = 0 \\[2mm] -2\dfrac{k}{R}V(y_0 - Vt) - 2kv_{\mathrm{rad}}, & v_{\mathrm{rad}} \neq 0 \end{cases} \tag{6.4}$$

Again consider Eqns (1.13) and (6.3), and notice that in both the cases due to the matched filtration, the signals returned from the point scatterer focus in the image plane in the vicinity of the point with zero Doppler frequency. It means that the images of two nearby point scatterers having different radial velocities can be spaced apart rather far, specifically over the distance $(R/V)\Delta v_{\mathrm{rad}}$, where Δv_{rad} is the difference between the radial velocities of the two scatterers. At the same time, the images of two or more spaced-apart scatterers can turn out to overlap in the image plane; this effect, as we shall see later, plays a significant role in SAR imaging of the ocean waves.

Let there be n point scatterers with variously valued and directed velocities as well as random and mutually non-correlated values of the reflection coefficient in the resolution cell. Then we substitute

$$p(y') = \sum_n a_n \delta(y'-y_n) \exp\left(-2ikv_{\mathrm{rad}}^{(n)} t'\right) \tag{6.5}$$

and obtain for the average intensity of SAR signal:

$$\langle I_{\mathrm{SAR}} \rangle = \sum_n \left\langle I_{\mathrm{SAR}}^{(n)} \right\rangle \tag{6.6}$$

where $I_{\mathrm{SAR}}^{(n)}$ is the intensity of signal reflected by a particular scatterer. As a result, the resolution cell shifts as a whole by $(R/V)\langle v_{\mathrm{rad}} \rangle$ (where $\langle v_{\mathrm{rad}} \rangle$ is the scatterer average radial velocity) and expands approximately by $\delta y = 2(R/V)\sigma_{rad}$, where σ_{rad} is the RMS of their orbital velocity.

Thus, SAR resolution cell expansion is due to the discrepancy in the velocities of the scatterers and its shift is accounted by the average velocity of the scatterers. In the context of roughness, it means the dependence of the resolution cell expansion and shift as a whole respectively on the small-scale against $\Delta_{0,SAR}$ (i.e. of sub-resolution scale) and large-scale components of roughness spectrum (or orbital velocity spectrum, to be more exact).

Certain expansion of the resolution cell is also induced by orbital accelerations. This effect can be examined as brought about by an alteration in the image shift due to the change δv_{rad} of the orbital velocity radial component during SAR integration time. According to Hasselmann et al. (1985), the corresponding RMS image smearing can be estimated as

$$\delta y = \frac{1}{2\sqrt{3}} \frac{R}{V} a_{rad} \Delta t \qquad (6.7)$$

where a_{rad} is the acceleration of the radial component of the orbital velocity. If we accept the estimate $a_{rad} \approx v_{rad}/T$, where T is the characteristic period of the large-scale roughness, and take into account that SAR integration time is set so that $\Delta t/T \ll 1$, then it becomes clear that the orbital acceleration effect is less significant than the orbital velocity impact.

Furthermore we will go into much more detail on the surface movement impact, and for the time being we will just mention that in this section we give only the overview of rough ocean surface SAR imaging mechanism. As a more profound analysis proves it, the mechanism most of the time is a much more complicated and essentially nonlinear.

6.2 SAR IMAGE CORRELATION FUNCTION: GENERAL RELATIONSHIPS

Consider expression (4.6) for the reflected field complex amplitude using Eqn. (4.5) as well as approximation (1.7):

$$a(x,t) = \frac{2k^2}{\sqrt{\pi}} e^{2ikR} \int_{\Delta\vec{r}} d\vec{r}' m(\vec{r}',t)\xi(\vec{r}',t) \exp\left\{ 2ik \left[\frac{(x'-x)\sin\theta_0}{-\zeta(\vec{r}',t)\cos\theta_0 + \frac{1}{2R}(y'-Vt)^2} \right] \right\} \qquad (6.8)$$

Unlike Eqn. (4.5) we have here the incidence angle θ_0, instead of the depression angle ψ_0, which is more common of space-based radar surveillance. Remember that the multiplier $m(\vec{r}',t)$ describes the backscattered signal modulation and includes both modulation types – tilt and hydrodynamic, and $\xi(t)$ is statistically homogeneous ("standard") ripples with steady characteristic features along the large wave profile.

Applying the aperture synthesis operation (6.1) to Eqn. (6.8), we obtain

$$a_{SAR}(x,t) = \frac{2k^2}{\Delta t \sqrt{\pi}} e^{2ikR} \int_{-\infty}^{\infty} dt' \Phi_t(t-t') \int_{\Delta \vec{r}} d\vec{r}' m(\vec{r}',t') \xi(\vec{r}',t')$$

$$\times \exp\left\{ 2ik \left[\begin{array}{c} (x'-x)\sin\theta_0 \\ -\zeta(\vec{r}',t')\cos\theta_0 + \frac{1}{2R}\left(y'^2 - 2Vt'(y'-Vt) - V^2t^2 \right) \end{array} \right] \right\}$$

(6.9)

The function $\Phi_t(t-t')$ essentially differs from zero only at $|t-t'| \leq \Delta t$, it is inserted into Eqn. (6.9) to level out the side lobes like $(\sin^2 u)/u^2$ in the SAR response to the point scatterer (see Eqn. (1.13)).

We introduce SAR signal intensity $I = a_{SAR} a_{SAR}^*$ and compose a corresponding correlation function:

$$B_I(\vec{\rho}) = \langle I(\vec{r}) I(\vec{r}+\vec{\rho}) \rangle$$

(6.10)

vector $\vec{\rho}$ has components ρ_x and $\rho_y = V\tau$, where τ is the temporal shift. Inserting into Eqn. (6.10) the equation for the intensity $I = a_{SAR} a_{SAR}^*$ where a_{SAR} is determined by Eqn. (6.9), we write

$$B_I = \frac{16k^8}{\pi^2 (\Delta t)^4} \iint_{\Delta t} dt_1\, dt_2\, \Phi_t(t-t_1)\Phi_t(t-t_2) \iint_{\Delta(t+\tau)} dt_3\, dt_4\, \Phi_t(t-t_3)\Phi_t(t-t_4)$$

$$\times \iint_{\Delta \vec{r}} d\vec{r}_1\, d\vec{r}_2 \iint_{\Delta(\vec{r}+\vec{\rho})} d\vec{r}_3\, d\vec{r}_4 \left\langle m(\vec{r},t_1)m^*(\vec{r}_2,t_2)m(\vec{r}_3,t_3)m^*(\vec{r}_4,t_4) \right.$$

$$\times \xi(\vec{r}_1,t_1)\xi(\vec{r}_2,t_2)\xi(\vec{r}_3,t_3)\xi(\vec{r}_4,t_4)$$
$$\times \exp\{2ik[(x_1 - x_2 + x_3 - x_4)\sin\theta_0$$
$$- [\zeta(\vec{r}_1,t_1) - \zeta(\vec{r}_2,t_2) + \zeta(\vec{r}_3,t_3) - \zeta(\vec{r}_4,t_4)]\cos\theta_0]\}$$
$$\times \exp\left\{ \frac{ik}{R}[y_1^2 - y_2^2 - 2V(t_1 y_1 - t_2 y_2) + 2V^2 t(t_1 - t_2)] \right\}$$
$$\left. \times \exp\left\{ \frac{ik}{R_+}[y_3^2 - y_4^2 - 2V(t_3 y_3 - t_4 y_4) + 2V^2(t+\tau)(t_3 - t_4)] \right\} \right\rangle$$

(6.11)

where $R_+ = R + \rho_x \sin\theta_0$; the symbols $\Delta(t+\tau)$ and $\Delta(\vec{r}+\vec{\rho})$ denote the areas shifted against Δt and $\Delta \vec{r}$ by τ and $\vec{\rho}$, respectively.

As done before (see Chapter 4), we take into account the statistical independence of "standard" ripples and large-scale roughness and average over the realizations of ripple normal field making use of this property (see e.g. Levin 1969):

$$\langle \xi(\vec{r}_1,t_1)\xi(\vec{r}_2,t_2)\xi(\vec{r}_3,t_3)\xi(\vec{r}_4,t_4) \rangle = B_\xi\left(\vec{r}_1 - \vec{r}_2, |t_1 - t_2|\right) B_\xi\left(\vec{r}_3 - \vec{r}_4, |t_3 - t_4|\right)$$
$$+ B_\xi\left(\vec{r}_1 - \vec{r}_4, |t_1 - t_4|\right) B_\xi\left(\vec{r}_2 - \vec{r}_3, |t_2 - t_3|\right) + B_\xi\left(\vec{r}_1 - \vec{r}_3, |t_1 - t_3|\right) B_\xi\left(\vec{r}_2 - \vec{r}_4, |t_2 - t_4|\right)$$

(6.12)

The transformation results in

$$B_I(\vec{\rho}) = B_1(\vec{\rho}) + B_2(\vec{\rho}) + B_3(\vec{\rho}) \tag{6.13}$$

$$B_1(\vec{\rho}) = \langle I_1(\vec{r})I_1(\vec{r}+\vec{\rho})\rangle \tag{6.14}$$

$$B_2(\vec{\rho}) = \langle I_2(\vec{r},\vec{\rho})I_2^*(\vec{r},\vec{\rho})\rangle \tag{6.15}$$

$$B_3(\vec{\rho}) = \langle I_3(\vec{r},\vec{\rho})I_3^*(\vec{r};\vec{\rho})\rangle \tag{6.16}$$

$$I_1 = \frac{4k^4}{\pi(\Delta t)^2}\iint dt_1 dt_2 \Phi_t(t-t_1)\Phi_t(t-t_2)\iint_{\Delta\vec{r}} d\vec{r}_1 d\vec{r}_2 m(\vec{r}_1,t_1)m^*(\vec{r}_2,t_2)B_\xi(\vec{r}_1-\vec{r}_2,t_1-t_2)$$
$$\times \exp\left\{2ik\left[\begin{array}{l}(x_1-x_2)\sin\theta_0 - \left[\zeta(\vec{r}_1,t_1)-\zeta(\vec{r}_2,t_2)\right]\cos\theta_0 \\ +\frac{1}{2R}\left[y_1^2-y_2^2-2V(t_1 y_1-t_2 y_2)+2V^2 t(t_1-t_2)\right]\end{array}\right]\right\} \tag{6.17}$$

$$I_2 = \frac{4k^4}{\pi(\Delta t)^2}\int dt_1 \Phi_t(t-t_1)\int dt_2 \Phi_t(t+\tau-t_2)\int_{\Delta\vec{r}} d\vec{r}_1$$
$$\times \int_{\Delta(\vec{r}+\vec{\rho})} d\vec{r}_2 m(\vec{r}_1,t_1)m^*(\vec{r}_2,t_2)B_\xi(\vec{r}_1-\vec{r}_2,t_1-t_2)$$
$$\times\exp\left\{2ik\left[\begin{array}{l}(x_1-x_2)\sin\theta_0 - \left[\zeta(\vec{r}_1,t_1)-\zeta(\vec{r}_2,t_2)\right]\cos\theta_0 \\ +\frac{1}{2R}\left[y_1^2-y_2^2-2V(t_1 y_1-t_2 y_2)+2V^2 t(t_1-t_2)-2V^2 t_2\tau\right]\end{array}\right]\right\} \tag{6.18}$$

$$I_3 = \frac{4k^4}{\pi(\Delta t)^2}\int dt_1 \Phi_t(t-t_1)\int dt_2 \Phi_t(t+\tau-t_2)\int_{\Delta\vec{r}} d\vec{r}_1$$
$$\times \int_{\Delta(\vec{r}+\vec{\rho})} d\vec{r}_2 m(\vec{r}_1,t_1)m^*(r_2,t_2)B_\xi(\vec{r}_1-\vec{r}_2,t_1-t_2)$$
$$\times \exp\left\{2ik\left[\begin{array}{l}(x_1+x_2)\sin\theta_0 - \left[\zeta(\vec{r}_1,t_1)+\zeta(\vec{r}_2,t_2)\right]\cos\theta_0 \\ +\frac{1}{2R}\left[y_1^2+y_2^2-2V(t_1 y_1+t_2 y_2)+2V^2 t(t_1+t_2)+2V^2 t_2\tau\right]\end{array}\right]\right\} \tag{6.19}$$

Remember that B_ξ is the spatio-temporal correlation function of "standard" ripples. Hereafter, for simplicity, we will discard the modulus in the argument $|t_1-t_2|$ of the correlation function B_ξ. When computing Eqns (6.18) and (6.19) we, as earlier (see Chapter 4), replaced

R_+ by R. Besides, indices 1 and 4 that, according to Eqns (4.9) and (4.10), should be present in Eqn. (6.18), and also 1 and 3 in Eqn. (6.19) are replaced by 1 and 2.

Integrals I_1, I_2, I_3 are the random field $\zeta(\vec{r}, t)$ functionals, and therefore averaging in Eqns (6.14)–(6.16) is over the large-scale rough sea realizations. Notably, these integrals do not sum to SAR signal intensity, just as B_I does not equal the sum of correlation functions for I_1, I_2, I_3. The relation between Eqns (6.14) and (6.16) indicates that out of the three summands from the right-hand side of Eqn. (6.13), only B_1 is the correlation function of the real (as proved below) value I_1.

Later we will see that the summands B_1 and B_2 of the correlation function B_I describe the roughness image proper and speckle underlayer, respectively. As for B_3, due to the presence of fast oscillating multipliers in the integrand of I_3, B_3 ends up to have a smaller value than B_2, at least by $(\lambda/\Delta x)^2$ times; therefore we can ignore the last summand on the right-hand side of Eqn. (6.13).

In the next two sections we analyse the integrals I_1, I_2 and obtain physically transparent expressions for them. For this purpose, we introduce new variables

$$\vec{r}_1 - \vec{r}_2 = \vec{\rho}', \quad \vec{r}_1 + \vec{r}_2 = 2\,\vec{r}' \tag{6.20a}$$

$$t_1 - t_2 = \tau', \quad t_1 + t_2 = 2t' \tag{6.20b}$$

As the function $B_\xi(\vec{\rho}', \tau')$ decreases fast at $\rho' > \rho_\xi, \tau' > \tau_\xi$, where ρ_ξ and τ_ξ are spatial and temporal ripple correlation scales, respectively, the integrands in I_1 and I_2 are essentially non-zero solely at $\rho' \leq \rho_\xi, \tau' \leq \tau_\xi$. Thus, keeping in mind the smallness of ρ_ξ and τ_ξ compared to the respective dimensions of large waves, we introduce

$$m(\vec{r}_1, t_1) = m(\vec{r}_2, t_2) = m(\vec{r}', t') \tag{6.21a}$$

$$\zeta(\vec{r}_1, t_1) - \zeta(\vec{r}_2, t_2) = \frac{\partial \zeta}{\partial x}(\vec{r}', t')\rho'_x + \frac{\partial \zeta}{\partial y}(\vec{r}', t')\rho'_y + \frac{\partial \zeta}{\partial t}(\vec{r}', t')\tau' \tag{6.21b}$$

where ρ'_x and ρ'_y are the components of the vector $\vec{\rho}'$. Finally, we take the range size Δx of the physical resolution cell, significant against ρ_ξ as small against the characteristic wavelength of large-scale roughness.

Thus, integrals I_1, I_2 through Eqns (6.14) and (6.15) describe the full correlation function of the water surface image.

6.3   INTENSITY OF THE SAR SIGNAL FORMING THE "IMAGE ITSELF"

We turn to the integral I_1, whose correlation function B_1 is that of the image proper, i.e. the image free of speckle noise. We take Φ_t as

$$\Phi_t(t - t_{1,2}) = \exp\left[-2\left(\frac{t - t_{1,2}}{\Delta t}\right)^2\right] \tag{6.22}$$

and write in view of the above remarks,

$$I_1 = \frac{4k^4}{\pi(\Delta t)^2} \Delta x \int\limits_{-\infty}^{\infty} \mathrm{d}y' \int\limits_{-\infty}^{\infty} \mathrm{d}t' \exp\left[-4\frac{(t-t')^2}{(\Delta t)^2}\right] \left|m(\vec{r}',t')\right|^2$$

$$\times \int\limits_{-\infty}^{\infty} \mathrm{d}\tau' \exp\left\{-\frac{\tau'^2}{(\Delta t)^2} - 2ik\left[\frac{\partial \zeta}{\partial t}(\vec{r}',t')\cos\theta_0 - \frac{V}{R}(Vt - y')\right]\tau'\right\}$$

$$\times \int\limits_{-\infty}^{\infty} \mathrm{d}\vec{\rho}' B_\xi(\vec{\rho}',\tau') \exp\left\{2ik\left[\begin{array}{l}\left(\sin\theta_0 - \frac{\partial\zeta}{\partial x}(\vec{r}',t')\cos\theta_0\right)\rho'_x \\ +\left(\frac{y'-Vt'}{R} - \frac{\partial\zeta}{\partial y}(\vec{r}',t')\cos\theta_0\right)\rho'_y\end{array}\right]\right\} \quad (6.23)$$

Integration over y' is formally applied to infinite limits, since as we shall see later, the main contribution to the integral is provided by relatively narrow stripe near $y' = Vt$, we followed the same pattern with the integral over $\vec{\rho}'$, where only the area $\rho' \leq \rho_\xi$ is meaningful (the "∞" symbol is omitted later with infinite limits of integration).

We set out the internal integral over $\vec{\rho}'$ in accordance with the Wiener–Khintchin theorem as follows:

$$\int \mathrm{d}\vec{\rho}' B_\xi(\vec{\rho}',\tau') \exp(-i\vec{\kappa}_{\mathrm{res}}\,\vec{\rho}') = 4\pi^2 \int \mathrm{d}\omega \Psi_\xi(\vec{\kappa}_{\mathrm{res}},\omega)\exp(-i\omega\tau') \quad (6.24)$$

where $\Psi_\xi\left(\vec{\kappa}_{\mathrm{res}},\omega\right)$ is the spatio-temporal ripple spectrum at the spatial frequency

$$\vec{\kappa}_{\mathrm{res}} = \left\{-2k\left(\sin\theta_0 - \frac{\partial\zeta}{\partial x}(\vec{r}',t')\cos\theta_0\right); \quad -2k\left(\frac{y'-Vt'}{R} - \frac{\partial\zeta}{\partial y}(\vec{r}',t')\cos\theta_0\right)\right\}$$

$$(6.25)$$

which is the resonance frequency of Bragg scatter. The presence of $\partial\zeta/\partial x$ and $\partial\zeta/\partial y$ reflects the fact that here $\vec{\kappa}_{\mathrm{res}}$ is the local spatial frequency dependent on the large-scale surface slopes at the current point $\vec{r}'$ in the incident plane and the surface perpendicular to it, and the term $(y'-Vt')/R$ reflects the dependence of the resonance frequency on the azimuthal angle, at which this point is probed by the antenna. Due to the insignificance of these allowances we shall suppose

$$\vec{\kappa}_{\mathrm{res}} \approx \{-2k\sin\theta_0; 0\} \quad (6.26)$$

The expression is equivalent to the "standard" ripples being situated on the plane surface, yet at each point $\vec{r}'$ having a respective orbital velocity induced by the large wave. Besides, the reflected signal intensity modulation is expressed by the multiplier $|m(\vec{r}',t')|^2$.

We assume for simplicity that the ripple spectrum lacks the components negatively projected on the wind direction. This means we take into account only Bragg ripple wave approaching or receding relatively to SAR look direction. Suppose for definiteness

$\vec{k}\vec{U} < 0$, where $\vec{U}$ is the wind speed vector, i.e. assuming the Bragg wave is propagating towards the radar, we obtain (see Eqn. (5.51))

$$\Psi_\xi\left(\vec{\kappa}_{\text{res}}, \omega\right) = \frac{1}{2} W_\xi(\vec{\kappa}_{\text{res}})\delta\left(\omega - \left(\Omega_{\text{res}} + \vec{\kappa}_{\text{res}}\vec{v}_{\text{orb}}\right)\right) \qquad (6.27)$$

where $W_\xi(\vec{\kappa}) = 2\hat{W}_\xi^+(\vec{\kappa})$ is the spatial (non-symmetrical) ripple spectrum and Ω is its own temporal frequency connected with $\vec{\kappa}$ by the dispersion correlation. Let us recall that $\hat{W}_\xi^+(\vec{\kappa})$ is half of the symmetrical spatial spectrum $\hat{W}_\xi(\vec{\kappa})$ corresponding to the positive temporal frequencies (see Chapter 2).

Inserting Eqn. (6.27) into (6.24), after we integrate over ω and substitute the result into Eqn. (6.23), we get

$$I_1 = \frac{8\pi k^4}{(\Delta t)^2} W_\xi(\vec{\kappa}_{\text{res}})\Delta x \int \mathrm{d}y' \int \mathrm{d}t' \exp\left[-4\frac{(t-t')^2}{(\Delta t)^2}\right]|m(x, y'; t'|^2$$

$$\times \int \mathrm{d}\tau' \exp\left\{-\frac{\tau'^2}{(\Delta t)^2} + 2ik\frac{V}{R}\left[(Vt - y') - \frac{R}{V}\left(v_{\text{rad}}(x, y'; t') + v_{\text{res}}^{\text{ph}}\sin\theta_0\right)\right]\tau'\right\} \qquad (6.28)$$

where v_{rad} is the radial component of the orbital velocity held positive when heading towards the radar. When writing Eqn. (6.28) we took into account that

$$2k\frac{\partial\zeta}{\partial t}\cos\theta_0 + \vec{\kappa}_{\text{res}}\vec{v}_{\text{orb}} = -2\vec{k}\vec{v}_{\text{orb}} = 2kv_{\text{rad}} \qquad (6.29)$$

The value $v_{\text{res}}^{\text{ph}} = \Omega_{\text{res}}/2k\sin\theta_0$ is the inherent phase velocity of resonance ripples. Furthermore, we shall not include $v_{\text{res}}^{\text{ph}}\sin\theta$ value as it does not significantly shift the image.

Integration over τ' is extremely easy. If we assume $\Delta t << T_0$, where T_0 is the inherent period of a large wave, we can also integrate over t'. In the end, we obtain

$$I_1(x, Vt) = \frac{\pi}{2}\Delta x \int \mathrm{d}y'\sigma_0(x, y')\exp\left\{-\frac{\pi^2}{\Delta_{0,\text{SAR}}^2}\left[Vt - y' - \frac{R}{V}v_{\text{rad}}(x, y')\right]\right\} \qquad (6.30)$$

As we see, significant contribution to the integral is made only by the area (or areas) with the azimuthal size of the SAR resolution cell order, and the resolution cell is located near the point where the exponent argument becomes zero, i.e. in the proximity of the intersection point of the straight line $Vt - y'$ and the random curve $(R/V)v_{\text{rad}}$.

Notably if we set the function Φ_t as

$$\Phi_t = \begin{cases} 1, & |t - t'| \leq \Delta t \\ 0, & |t - t'| > \Delta t \end{cases} \qquad (6.31)$$

then we would get

$$I_1 = \Delta x \int dy' \sigma_0(x, y') \left[\frac{\sin w(x, y')}{w(x, y')} \right]^2 \tag{6.32}$$

$$w = \frac{\pi}{\Delta_{0,\text{SAR}}} \left[Vt - y' - \frac{R}{V} v_{\text{rad}}(x, y') \right] \tag{6.33}$$

Later on, we will use the function Φ_t in the shape of either Eqn (6.22) or (6.31) governed by convenience in each particular case.

6.4 SPECTRUM OF THE SAR IMAGE OF THE OCEAN

In the present section we will examine SAR image spectrum, that is the image itself, i.e. the speckle-noise-free image described by Eqn. (6.30). It is to point out that the spectrum formula was first obtained in Hasselmann and Hasselmann (1991), which analysed an ideal case where $\Delta_{0,\text{SAR}} = 0$.

We rewrite Eqn. (6.30) replacing Vt by y:

$$I_1(x, y) = \frac{\pi}{2} \Delta x \int dy' \sigma_0(x, y') \exp\left\{ -\frac{\pi^2}{\Delta_{0,\text{SAR}}^2} \left[y - y' - \frac{R}{V} v_{\text{rad}}(x, y') \right]^2 \right\} \tag{6.34}$$

and then convert it into

$$I_1(x, y) = \frac{1}{4\pi^{1/2}} \Delta x \Delta_{0,\text{SAR}} \int dy' \sigma_0(x, y')$$
$$\times \int d\kappa_y \exp\left\{ -\frac{\Delta_{0,\text{SAR}}^2}{4\pi^2} \kappa_y^2 + i \left[y - y' - \frac{R}{V} v_{\text{rad}}(x, y') \right] \kappa_y \right\} \tag{6.35}$$

It is easy to establish that Eqns (6.34) and (6.35) are identical if we take into account the relation

$$\int dx \exp\left(-p^2 x^2 \pm iqx \right) = \frac{\sqrt{\pi}}{p} \exp\left(-\frac{q^2}{4p^2} \right) \tag{6.36}$$

Using Eqn. (6.35) we compose the correlation function

$$B_{1,\text{SAR}}(\rho_x, \rho_y) = \langle I_1(x, y) I_1(x + \rho_x, y + \rho_y) \rangle \tag{6.37}$$

and by analogy with Eqn. (5.54) we assume

$$\left\langle \sigma_0(x, y') \sigma_0(x + \rho_x, y'') \exp\left[i \frac{R}{V} v_{\text{rad}}(x, y') \kappa_y' \right] \exp\left[i \frac{R}{V} v_{\text{rad}}(x + \rho_x, y'') \kappa_y'' \right] \right\rangle$$
$$\approx \langle \sigma_0(x, y') \sigma_0(x + \rho_x, y'') \rangle \left\langle \exp\left[i \frac{R}{V} v_{\text{rad}}(x, y') \kappa_y' \right] \exp\left[i \frac{R}{V} v_{\text{rad}}(x + \rho_x, y'') \kappa_y'' \right] \right\rangle \tag{6.38}$$

Then

$$
B_{1,\text{SAR}} = \frac{1}{16\pi}(\Delta x)^2 \Delta_{0,\text{SAR}}^2 \iint dy' dy'' \langle \sigma_0(x,y') \sigma_0(x+\rho_x, y'') \rangle
$$
$$
\times \iint dk'_y\, dk''_y \exp\left[-\frac{\Delta_{0,\text{SAR}}^2}{4\pi^2} \left(\kappa'^2_y + \kappa''^2_y \right) \right]
$$
$$
\times \exp\left[i(y - y')\kappa'_y + i(y + \rho_y - y'')\kappa''_y \right]
$$
$$
\times \left\langle \exp\left[i\frac{R}{V} v_{\text{rad}}(x,y')\kappa'_y \right] \exp\left[i\frac{R}{V} v_{\text{rad}}(x+\rho_x, y'')\kappa''_y \right] \right\rangle \tag{6.39}
$$

Evidently, the first multiplier of integrand in Eqn. (6.39) having angular brackets is the correlation function of radar cross section, and we will term the correlation function of the RAR signal, B_{RAR}, following Alpers et al. (1981). (Therefore we have used subscript SAR in the expression for B_1.) The second multiplier is nothing but the characteristic function of two variables, Θ_2. As clearly seen, both these multipliers depend on the distance between the points (x,y') and $(x + \rho_x, y'')$; therefore after we introduce the new variable $\rho'_y = y'' - y'$, we can write Eqn. (6.39) as

$$
B_{1,\text{SAR}} = \frac{1}{16\pi}(\Delta x)^2 \Delta_{0,\text{SAR}}^2 \iint d\kappa'_y d\kappa''_y \exp\left(i\kappa''_y \rho_y \right) \exp\left[-\frac{\Delta_{0,\text{SAR}}^2}{4\pi^2} \left(\kappa'^2_y + \kappa''^2_y \right) \right]
$$
$$
\times \int d\rho'_y \exp\left(-i\kappa''_y \rho'_y \right) B_{\text{RAR}}\left(\rho_x, \rho'_y \right) \Theta_2\left(\rho_x, \rho'_y \right) \int dy' \exp\left[i(y - y') \left(\kappa'_y + \kappa''_y \right) \right] \tag{6.40}
$$

Since

$$
\int dy' \exp\left[i(y - y')\left(\kappa'_y + \kappa''_y \right) \right] = 2\pi \delta\left(\kappa'_y + \kappa''_y \right) \tag{6.41}
$$

after we integrate over κ''_y, we obtain

$$
B_{1,\text{SAR}} = \frac{1}{8}(\Delta x)^2 \Delta_{0,\text{SAR}}^2 \int d\kappa'_y \exp\left(-i\kappa'_y \rho_y \right) \exp\left[-\frac{\Delta_{0,\text{SAR}}^2}{2\pi^2} \kappa'^2_y \right]
$$
$$
\times \int d\rho'_y \exp\left(i\kappa'_y \rho'_y \right) B_{\text{RAR}}\left(\rho_x, \rho'_y \right) \Theta_2\left(\rho_x, \rho'_y \right) \tag{6.42}
$$

where

$$
\Theta_2 = \left\langle \exp\left[i\frac{R}{V} v_{\text{rad}}(x,y')\kappa'_y \right] \exp\left[-i\frac{R}{V} v_{\text{rad}}\left(x+\rho_x, y'+\rho'_y\right)\kappa'_y \right] \right\rangle \tag{6.43}
$$

For the Gaussian distribution of the random field v_{rad}, the following formula is valid:

$$
\Theta_2 = \exp\left\{ -\left(\frac{R}{V} \right)^2 \sigma_{\text{rad}}^2 \left[1 - r_{\text{rad}}\left(\rho_x, \rho'_y \right) \right] \kappa'^2_y \right\} \tag{6.44}
$$

where σ_{rad}^2 is the variance and r_{rad} the normalized correlation function (correlation coefficient) of v_{rad}. Thus,

$$\hat{B}_{\text{SAR}} \propto \int d\kappa'_y \exp\left(-i\kappa'_y \rho_y\right) \exp\left\{ -\left[\frac{\Delta_{0,\text{SAR}}^2}{2\pi^2} + \left(\frac{R}{V}\right)^2 \sigma_{\text{rad}}^2 \right] \kappa_y'^2 \right\}$$
$$\times \int d\rho'_y \exp\left(i\kappa'_y \rho_y'\right) B_{\text{RAR}}\left(\rho_x, \rho_y'\right) \exp\left[\left(\frac{R}{V}\right)^2 \sigma_{\text{rad}}^2 r_{\text{rad}}\left(\rho_x, \rho_y'\right) \kappa_y'^2 \right] \quad (6.45)$$

We set the last exponential multiplier in Eqn. (6.45) as a series:

$$\exp\left[\left(\frac{R}{V}\right)^2 \sigma_{\text{rad}}^2 r_{\text{rad}}(\rho_x, \rho_y') \kappa_y'^2 \right] = 1 + \sum_{n=1}^{\infty} \frac{1}{n!} \left(\frac{R}{V}\right)^{2n} \sigma_{\text{rad}}^{2n} r_{\text{rad}}^n(\rho_x, \rho_y') \kappa_y'^{2n} \quad (6.46)$$

and consider that

$$B_{\text{RAR}} = \langle \sigma_0 \rangle^2 + b_{\text{RAR}}(\rho_x, \rho_y') \quad (6.47)$$

Then

$$B_{1,\text{SAR}} \propto B_{1,\text{SAR}}^0 + \int d\kappa'_y \exp\left(-i\kappa'_y \rho_y\right) \exp\left\{ -\left[\frac{\Delta_{0,\text{SAR}}^2}{2\pi^2} + \left(\frac{R}{V}\right)^2 \sigma_{\text{rad}}^2 \right] \kappa_y'^2 \right\} \quad (6.48)$$
$$\times \left\{ \int d\rho'_y \exp\left(i\kappa'_y \rho_y'\right) b_{\text{RAR}}(\rho_x, \rho_y') + \sum_{n=1}^{\infty} \frac{1}{n!} \left(\frac{R}{V}\right)^{2n} \right.$$
$$\left. \times \sigma_{\text{rad}}^{2n} \kappa_y'^{2n} \int d\rho'_y \exp\left(i\kappa'_y \rho_y'\right) B_{\text{RAR}}(\rho_x, \rho_y') r_{\text{rad}}^n(\rho_x, \rho_y') \right\}$$

where $B_{1,\text{SAR}}^0 \propto \langle \sigma_0 \rangle^2$ is the constant component. We subtract the constant component from $B_{1,\text{SAR}}$, multiply the remaining part by $(2\pi)^{-2} \exp(-i\vec{\kappa}\vec{\rho})$ and integrate first over ρ_y and then κ'_y. As a result, the expression for the image spectrum is given by

$$\hat{W}_{1,\text{SAR}}(\vec{\kappa}) \propto \exp\left\{ -\left[\frac{\Delta_{0,\text{SAR}}^2}{2\pi^2} + \left(\frac{R}{V}\right)^2 \sigma_{\text{rad}}^2 \right] \kappa_y^2 \right\}$$
$$\times \left\{ \hat{W}_{\text{RAR}}(\vec{\kappa}) + (2\pi)^{-2} \sum_{n=1}^{\infty} \frac{1}{n!} \left(\frac{R}{V}\right)^{2n} \sigma_{\text{rad}}^{2n} \kappa_y^{2n} \int d\vec{\rho} \exp(-i\vec{\kappa}\vec{\rho}) B_{\text{RAR}}(\vec{\rho}) r_{\text{rad}}^n(\vec{\rho}) \right\} \quad (6.49)$$

where

$$\hat{W}_{\text{RAR}}(\vec{\kappa}) = \frac{1}{4\pi^2} \int d\vec{\rho}\, b_{\text{RAR}}(\vec{\rho}) \exp(-i\vec{\kappa}\vec{\rho}) \quad (6.50)$$

is the spatial spectrum of radar cross-section fluctuations.

We see from Eqn. (6.49) that but for the surface motion, the SAR image spectrum would look as if it has been obtained with the help of RAR having the azimuthal resolution $\Delta_{0,\text{SAR}}$, i.e.

$$\hat{W}_{1,\text{SAR}}(\vec{\kappa}) \propto \hat{W}_{\text{RAR}}(\vec{\kappa}) \exp\left(-\frac{\Delta_{0,\text{SAR}}^2}{2\pi^2} \kappa_y^2 \right) \qquad (6.51)$$

However, the presence of the orbital velocities changes the situation significantly. What are the changes then? On the one hand, the azimuthal wave number ascending power series add high frequency components to the image spectrum. Yet, at rather high values of the ratio R/V characteristic for satellite-borne radar probing, the exponential multiplier as a rule has a prevailing influence, and the image spectrum becomes truncated at azimuthal wave number.

Because of the spectral cut-off, part of roughness spectrum at high wave number side becomes lost in the SAR image and consequently induces the shift of the spectrum peak to the low wave number side. Speaking about the peak shift we mean the azimuthal wave number shift, so in case of non-azimuthally directed waves the radius vector of the spectrum peak shortens and turns towards the range direction. As Eqn. (6.49) shows, along κ_y the image spectrum decreases by e times at

$$\kappa_y^{(0)} \approx \kappa_{\text{cut-off}} = \frac{R}{V} \sigma_{\text{rad}} \qquad (6.52)$$

The approximate equation sign implies that the cut-off factor contains also a term describing the SAR resolution cell influence. However, since as a rule

$$\Delta_{0,\text{SAR}} << 2\pi^2 \frac{R}{V} \sigma_{\text{rad}} \qquad (6.53)$$

the respective term in the cut-off factor can be dealt without.

The spectral peak shift effect is clearly demonstrated in Figure 6.1 from Vesecky and Stewart (1982). This figure displays the ocean image spectrum collected from SAR borne by the American oceanographic SEASAT satellite. More exactly, this is the spectrum section averaged over directions near (within about $\pm 15°$) the dominant wave direction. Alongside the SAR image spectrum, in Figure 6.1 we also see omnidirectional wave height and wave slope spectra as measured by pitch-and-roll buoy. Notably, under the image spectrum itself we see a "pedestal" of the speckle-noise spectrum; we will discuss it later.

Based on the information above and the experimental data, we can formulate the visual perception criteria for surface waves in SAR images of the ocean.

First, according to the resonance scatter theory, microwave radar sees roughness only if there are short-wave ripples on the surface with wavelengths fitting Bragg relation (see Section 3.2.2). This means the near-surface wind speed must exceed the threshold, which is approximately $2 - 3\,\text{m}\,\text{s}^{-1}$.

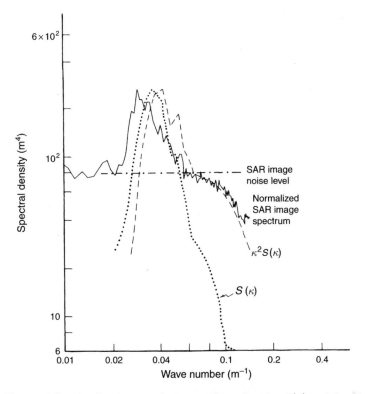

Figure 6.1 The omnidirectional spectrum of ocean surface elevation $S(\kappa)$ and the surface slope spectrum $\kappa^2 S(\kappa)$ compared with the spectrum of SAR image intensity. The spectra are normalized along the ordinate so that the peaks have the same values (Vesecky and Stewart 1982).

Then if we accept an azimuthally travelling wave as a visual perception criterion,

$$\Lambda^{az} \geq \Lambda^{az}_{min} = \frac{2\pi}{2\kappa_{cut\text{-}off}} = \pi \frac{R}{V} \sigma_{rad} \qquad (6.54)$$

then we obtain for the windsea conditions, taking into account Eqn. (2.23),

$$\Lambda^{az} \geq \Lambda^{az}_{min} \approx 0.2 \frac{R}{V} U \cos \theta_0 \qquad (6.55)$$

Considering Eqns (2.22) and (2.28) we find that

$$\Lambda^{az} \geq C_0 \frac{R}{V} \sqrt{H_s} \cos \theta_0 \qquad (6.56)$$

where $C_0 = 1.3\,\mathrm{m}^{1/2}\,\mathrm{s}^{-1}$. Criterion (6.56) in fact is identical to semi-empirical relationship for the minimum detectable azimuth wavelength cited in Beal et al. 1981 and Holt (2004).

As we can see from Eqn. (6.56), the wavelength visual perception threshold for azimuthally travelling waves grows as H_s increases. In contrast, the wave height should be sufficiently large for the roughness image to be discernible at the noise background caused by speckle noise as well as thermal noise originated in the SAR system itself. The experimental data presented in Stewart (1985) indicate the H_s threshold is approximately 1.4 m. Thus, as for the ability of SAR to see azimuthally travelling waves, the last two parameters limit the wave height at both ends.

It is clear from what has been said that if the wind wave spectrum has maximum at wave number $\vec{\kappa}_{\mathrm{max}} = \{\kappa_{\mathrm{max}} \cos \phi_0, \kappa_{\mathrm{max}} \sin \phi_0\}$ and $\kappa_{\mathrm{max}} \sin \phi_0 > 2\pi / \Lambda_y^{\mathrm{max}}$, where

$$\Lambda_y^{\mathrm{max}} = C_0 \frac{R}{V} \sqrt{H_s} \left(\cos^2 \theta_0 + \sin^2 \phi_0 \sin^2 \theta_0 \right)^{1/2} \tag{6.57}$$

then in the SAR image sea roughness looks as almost range directed, and the apparent dominant wavelength turns out significantly larger than the true one.

6.5   SAR IMAGING MECHANISMS

Equations (6.34) and (6.49) provide a formal mathematical description of SAR signal imaging part and the respective spectrum. However, due to the complex non-linear relation between roughness characteristics, on the one hand, and SAR image features, on the other, these equations need a thorough analysis to pinpoint the physical mechanisms governing SAR imaging of the ocean.

We conclude from above that one of these mechanisms is common for both SAR and RAR, namely, the fluctuations of radar cross section. As for the impact of surface motion (velocity bunching effect), we discussed in the previous section, we are yet to investigate the mechanisms responsible for the impact.

6.5.1   Sub-resolution velocities impact

Recall that in Section 6.1 guided by qualitative preliminary estimates, we mentioned that the orbital velocity fluctuations that are small scale compared to SAR nominal resolution cell lead to this very resolution cell smearing. To see how the outlined theory describes the phenomenon we consider Eqn. (6.30).

Notably, the function

$$f(w) = \exp(-w^2) \tag{6.58}$$

$$w = \frac{\pi}{\Delta_{0,\mathrm{SAR}}} \left[Vt - y' - \frac{R}{V} v_{\mathrm{rad}}(x, y') \right]$$

in the integrand of Eqn. (6.30) is essentially non-zero only if the straight line $Vt - y'$ and the random curve $(R/V)v_{rad}$ are in a significant vicinity to each other. Figure 6.2 displays an area containing points of intersection between the straight line and the random curve (a), and the function $f(w)$ in this area (b). The size of the interval, where $f(w)$ considerably decreases as it is receding from the intersection point where $w_0 = 0$, is of the order $\Delta_{0,SAR}$. Evidently, if two adjoining intersection points are spaced apart less than by $\Delta_{0,SAR}$, the function $f(w)$ between these points does not decrease; instead it stays close to unity. This exactly causes SAR resolution cell smearing. In Figure 6.2, vertical dotted lines indicate the area where the line and the random curve approach each other, and consequently the function $f(w)$ is close to unity – this is SAR-smeared resolution cell.

To give a quantitative estimate of this effect, we represent the orbital velocity as a sum of two components:

$$v_{orb} = \widehat{v}_{orb} + \tilde{v}_{orb} \tag{6.59}$$

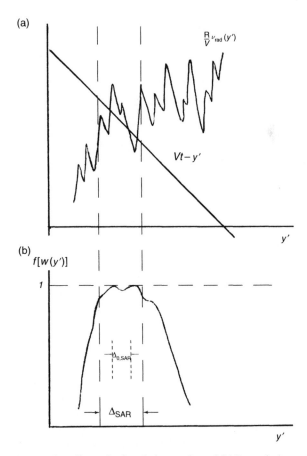

Figure 6.2 Illustration on the effect of azimuthal smearing of SAR resolution cell.

where $\widehat{v}_{\text{orb}}$ and $\tilde{v}_{\text{orb}}$ are large- and small-scale parts as compared to $\Delta_{0,\text{SAR}}$. We average the function $I_1(x, y)$, describing the intensity of SAR signal forming the SAR image itself, over sub-resolution radial velocities $\tilde{v}_{\text{rad}}$ assuming their normal distribution with zero mean and variance $\tilde{\sigma}_{\text{rad}}^2$.

As earlier, we assume

$$\langle \sigma_0 \exp(-w^2) \rangle \approx \langle \sigma_0 \rangle \langle \exp(-w^2) \rangle \tag{6.60}$$

Multiplying $f(w) = \exp(-w^2)$ by the Gaussian PDF and integrating over $\tilde{v}_{\text{rad}}$ in infinite limits, we get

$$\langle \exp(-w^2) \rangle = \frac{\Delta_{0,\text{SAR}}}{\Delta_{\text{SAR}}} \exp\left(-\widehat{w}^2\right) \tag{6.61a}$$

$$\widehat{w} = \frac{\pi}{\Delta_{\text{SAR}}} \left(Vt - y' - \frac{R}{V}\widehat{v}_{\text{rad}}\right) \tag{6.61b}$$

$$\Delta_{\text{SAR}} = \left[\Delta_{0,\text{SAR}}^2 + 2\pi^2 \left(\frac{R}{V}\right)^2 \tilde{\sigma}_{\text{rad}}^2\right]^{1/2} \tag{6.61c}$$

Therefore, the intensity $\widehat{I}_{1,\text{SAR}}$ of the signal forming SAR image itself averaged over sub-resolution orbital velocities is

$$\widehat{I}_{1,\text{SAR}}(x, y) = \frac{\pi}{2}\Delta x \frac{\Delta_{0,\text{SAR}}}{\Delta_{\text{SAR}}} \int dy' \widehat{\sigma}_0(x, y') \exp\left\{-\frac{\pi^2}{\Delta_{\text{SAR}}^2}\left[y - y' - \frac{R}{V}\widehat{v}_{\text{rad}}(x, y')\right]^2\right\} \tag{6.62}$$

where Δ_{SAR} is the SAR resolution cell smeared because of sub-resolution velocities on the ocean surface. (Anticipating further conclusions, we shall remark that as the mechanism of SAR imaging of the ocean waves is very much nonlinear, the role and significance of Δ_{SAR} linear characteristics are not in fact that apparent, as we could expect at first glance.)

In Alpers and Brüning (1986) the authors have assessed the value $(R/V)^2 \tilde{\sigma}_{\text{rad}}^2$ for SAR with SEASAT SAR parameters; the assessment results show that the azimuthal size of smeared resolution cell can exceed the nominal value by several times.

Let us examine the concept of scene coherence time τ_s; we introduce it with the help of the equation (see in Alpers and Brüning 1986)

$$\frac{2\pi^2 (R/V)^2 \tilde{\sigma}_{\text{rad}}^2}{\Delta_{0,\text{SAR}}^2} = \left(\frac{\Delta t}{\tau_s}\right)^2 \tag{6.63}$$

where Δt is the SAR integration time, and the numerator on the left-hand side according to Eqn. (6.61c) is increment to $\Delta_{0,\text{SAR}}^2$ due to smearing.

The scene coherence time can be understood as the period when the aperture can be synthesized without any significant loss of azimuthal resolution. At first glance, τ_s should have the value of the order of the "lifetime" of resonance (Bragg) roughness component. From this point of view it is unclear as to why then in case of SEASAT SAR the estimates mentioned above display considerable smearing of the resolution cell, although the integration time for SEASAT SAR is set less than the "lifetime" of roughness resonance component, which is several seconds for SEASAT SAR.

Equation (6.63) gives

$$\tau_s = \frac{\lambda}{2\sqrt{2\pi}\tilde{\sigma}_{rad}} \tag{6.64}$$

The computation performed in Alpers and Brüning (1986) for SEASAT SAR nominal resolution cell $25 \times 25 \, m^2$ and full developed windsea showed

$$\tau_s = \begin{cases} 0.14s, & U = 3 \, m \, s^{-1} \\ 0.047s, & U > 13 \, m \, s^{-1} \end{cases} \tag{6.65}$$

As we see, the scene coherence time turned out significantly less than the "lifetime" of roughness spectrum Bragg component, which is quite logical. These values could only be equal if the orbital velocity remained steady along the whole area of SAR nominal resolution cell. However, in situ when on the surface there are small-scale orbital velocities, τ_s is nothing but the correlation time of the radar signal backscattered by the surface with rms of $v_{orb.\,rad}$ equal to $\tilde{\sigma}_{rad}$. We can easily check it with the help of Eqn. (4.34) obtained in Chapter 4 for the backscattered signal correlation function.

We consider Eqn. (6.49) and write the cut-off factor separately:

$$F_{cut-off} = \exp\left\{ -\left[\frac{\Delta_{0,SAR}^2}{2\pi^2} + \left(\frac{R}{V}\right)^2 \sigma_{rad}^2 \right] \kappa_y^2 \right\} \tag{6.66}$$

By substituting into Eqn. (6.66)

$$\Delta_{0,SAR}^2 = \Delta_{SAR}^2 - 2\pi^2 \left(\frac{R}{V}\right)^2 \tilde{\sigma}_{rad}^2 \tag{6.67}$$

(see Eqn. (6.61c)), taking into account that $\sigma_{rad}^2 = \hat{\sigma}_{rad}^2 + \tilde{\sigma}_{rad}^2$, we get

$$F_{cut-off} = \exp\left\{ -\left[\frac{\Delta_{SAR}^2}{2\pi^2} + \left(\frac{R}{V}\right)^2 \hat{\sigma}_{rad}^2 \right] \kappa_y^2 \right\} \tag{6.68}$$

The cut-off factor acts as a low-pass filter for azimuthal wave numbers. This filter characteristics depends on both sub-resolution and large-scale orbital velocities. Therefore, spectral cut-off is not only the cause of image linear filtration by the smeared resolution cell, and, consequently, SAR imaging of the ocean is governed by an unknown so far and non-linear process (or processes).

6.5.2 Velocity bunching mechanism: linear and quasi-linear approximations

To proceed with the analysis we start off with the initial formula for intensity of speckle-noise-free SAR signal averaged over sub-resolution scales:

$$\widehat{I}_i(x, y = Vt) = \frac{\pi}{2} \Delta x \frac{\Delta_{0,\,\text{SAR}}}{\Delta_{\text{SAR}}} \int dy' \widehat{\sigma}_0(y') \exp\left\{ -\frac{\pi^2}{\Delta_{\text{SAR}}^2} \left[y - y' - \frac{R}{V} \widehat{v}_{\text{rad}}(x, y') \right] \right\} \quad (6.69)$$

This expression describing the "image itself" (subscript "i" – for "image") contains the smeared resolution cell and the orbital velocity smoothed over sub-resolution scales. Note that Eqn. (6.63) practically coincides with the equation for SAR signal intensity obtained by Alpers and Rufenach (1979).

We locate the x coordinate and examine the image line along SAR carrier flight direction, i.e. along the Y-axis.

The main contribution into integral (6.69) is by the zero neighbourhood of the exponential function argument in the integrand, i.e. the neighbourhood of the intersection points between the straight line $y - y'$ and the random function $(R/V)\widehat{v}_{\text{rad}}(y')$. As we shall further see, the number of these points significantly impacts the imagery.

It is geometrically evident that with the probability close to unity the equation

$$y - y' = \frac{R}{V} \widehat{v}_{\text{rad}}(y') \quad (6.70)$$

has a sole solution if the following relation is satisfied:

$$\frac{R}{V} \widehat{\sigma}_{\text{rad}} \ll \frac{\Lambda_v}{|\cos \phi_0|} \quad (6.71)$$

where Λ_v is a characteristic wavelength in the spectrum of the large-scale orbital velocity $\widehat{v}_{\text{orb}}$, and $\widehat{\sigma}_{\text{rad}}$ is the mean square root of this velocity radial component; ϕ_0 is the angle between the general wave propagation direction and SAR platform flight direction. In this case we expand $\widehat{v}_{\text{rad}}(y')$ into a series in the vicinity of the intersection point y_0 and using only the linear member of decomposition we get

$$\widehat{I}_i \propto \frac{\widehat{\sigma}_0(y_0)}{\left| 1 + \dfrac{R}{V} \dfrac{\partial \widehat{v}_{\text{rad}}}{\partial y}(y_0) \right|} \quad (6.72)$$

Computing Eqn. (6.72) we ignored the variation of the radar cross section within SAR resolution cell, which would have been unacceptable had we dealt with non-smoothed values. Consequently, in reality, i.e. at $\Delta_{0,\,\text{SAR}} \neq 0$, expression (6.72) works only for the averaged signal.

As Eqn. (6.72) clearly shows, SAR specifics manifest in the following way. First, SAR signal intensity at point y is proportional to the intensity of the field backscattered by the point with $y_0 = y + (R/V)\widehat{v}_{\text{rad}}(y_0)$ azimuthal coordinate non-coincident with $y = Vt$ (recall the image shift of the moving scatterer in Section 6.1). Secondly, Eqn. (6.72) has a multiplier independent of backscatter cross section but defined by orbital velocity field. This particularly means that SAR unlike RAR has the ability to see roughness even when there are no cross-section fluctuations – this is the case when probing is effected along surface wavefronts, i.e. when they are propagating azimuthally.

The denominator on the right-hand side of Eqn. (6.72) is different from unity because of the derivative $\partial\widehat{v}_{\text{rad}}/\partial y$, and its fluctuations bring about a change in the width of the maximum in the exponential function under integral in Eqn. (6.69). This means the fluctuations of the azimuthal size of the surface patch backscattering SAR signal. However, if we stick to the concept of the resolution cell having fixed azimuthal size, we can think of fluctuations of intensity $\widehat{I}_{\text{i}}$ generated by velocity field as induced by the equivalent fluctuations of surface scatterer effective density (Alpers and Rufenach 1979, Alpers et al. 1981). As $\partial\widehat{v}_{\text{rad}}/\partial y$ is in the denominator of formula (6.66), this velocity bunching mechanism, generally speaking, is the non-linear one.
If the following inequation is satisfied,

$$\left|\frac{R}{V}\frac{\partial\widehat{v}_{\text{rad}}}{\partial y}\right| \leq 0.3 \tag{6.73}$$

expression (6.72) can be rewritten as (Alpers and Rufenach 1979)

$$\widehat{I}_{\text{i}} \propto \widehat{\sigma}_0\left(1 - \frac{R}{V}\frac{\partial\widehat{v}_{\text{rad}}}{\partial y}\right) \tag{6.74}$$

Then we divide $\widehat{\sigma}_0$ into the constant $(\langle\widehat{\sigma}_0\rangle)$ and fluctuational $(\widehat{\sigma}'_0)$ parts and leaving only the linear members over the surface elevations, receive for the fluctuational part of the signal:

$$\widehat{I}'_{\text{i}} \propto \widehat{\sigma}'_0 - \langle\widehat{\sigma}_0\rangle\frac{R}{V}\frac{\partial\widehat{v}_{\text{rad}}}{\partial y} \tag{6.75}$$

Thus, if both Eqns (6.71) and (6.73) are satisfied, there is a linear relation between intensity fluctuations of SAR signal, responsible for rough sea imaging, on the one hand, and surface elevations ζ, on the other, i.e. SAR images roughness linearly (notethat we consider σ_0 and ζ values to be in linear relation). Expression (6.75) yields a formula for linear MTF:

$$T_{\text{SAR}}^{\text{lin}}(\vec{\kappa}) = T_{\text{RAR}}(\vec{\kappa}) + T_{\text{vb}}(\vec{\kappa}) \tag{6.76}$$

where in accordance with Eqn. (4.27)

$$T_{\text{vb}}(\vec{\kappa}) = \frac{R}{V}\left(\cos\theta_0 - i\frac{\kappa_x}{\kappa}\sin\theta_0\right)\sqrt{g\kappa}\kappa_y \tag{6.77}$$

is velocity bunching part of the linear MTF, and thus in the linear approximation

$$\widehat{I}_{i,\text{SAR}} - \langle \widehat{I}_{i,\text{SAR}} \rangle = \langle \widehat{\sigma}_0 \rangle \int d\vec{\kappa} T_{\text{SAR}}^{\text{lin}}(\vec{\kappa}) A_\zeta(\vec{\kappa}) \exp\left[i\left(\vec{\kappa}\vec{r} - \Omega(\kappa)t \right) \right] + \text{c.c.} \qquad (6.78)$$

where $A_\zeta(\vec{\kappa})$ is decomposition spectral amplitude

$$\zeta(\vec{r},t) = \int d\vec{\kappa} A_\zeta(\vec{\kappa}) \exp\left[i\left(\vec{\kappa}\vec{r} - \Omega(\kappa)t \right) \right] + \text{c.c.} \qquad (6.79)$$

With the help of Eqn. (6.78), the formula for the spectrum $\widehat{W}_{i,\text{SAR}}^{\text{lin}}$ is given by

$$\widehat{W}_{i,\text{SAR}}^{\text{lin}}(\vec{\kappa}) = \langle \widehat{\sigma}_0 \rangle^2 \left| T_{\text{SAR}}^{\text{lin}}(\vec{\kappa}) \right|^2 \widehat{W}_\zeta(\vec{\kappa}) \qquad (6.80)$$

Let us take a closer look at value $\left| T_{\text{SAR}}^{\text{lin}}(\vec{\kappa}) \right|^2$:

$$\left| T_{\text{SAR}}^{\text{lin}}(\vec{\kappa}) \right|^2 = \left| T_{\text{RAR}}(\vec{\kappa}) \right|^2 + \left| T_{\text{vb}}(\vec{\kappa}) \right|^2 + 2\left| T_{\text{RAR}}(\vec{\kappa}) \right| \left| T_{\text{vb}}(\vec{\kappa}) \right| \cos(\eta_{\text{RAR}} - \eta_{\text{vb}}) \qquad (6.81)$$

where η_{RAR} and η_{vb} are the phases of two parts of linear SAR MTS, and write the relation

$$C(\phi_0) = \frac{2\left| T_{\text{RAR}}(\vec{\kappa}) \right| \left| T_{\text{vb}}(\vec{\kappa}) \right|}{\left| T_{\text{RAR}}(\vec{\kappa}) \right|^2 + \left| T_{\text{vb}}(\vec{\kappa}) \right|^2} \cos(\eta_{\text{RAR}} - \eta_{\text{vb}}) = A(\phi_0) \cos \Delta\eta(\phi_0) \qquad (6.82)$$

where ϕ_0 is the angle between the wave propagation direction and the direction of SAR platform flight.

If we first analyse the interval $0° \leq \phi_0 \leq 180°$, one can see that $A(\phi_0)$ is null at $\phi_0 = 0°, 90°, 180°$ and has two maxima symmetrical against the point $\phi_0 = 90°$ and equal to unity. The accurate positions of maxima is defined by the relation between the maximum values of $|T_{\text{RAR}}|$ and $|T_{\text{vb}}|$ as functions of ϕ_0, attained respectively at $\phi_0 = 90°$ and at $\phi_0 = 0°, 180°$. Since $|T_{\text{vb}}|_{\text{max}} > |T_{\text{RAR}}|_{\text{max}}$ (see Figure 6.3 reproduced from Alpers et al. 1981), we conclude that the maxima at the intersection of curves $|T_{\text{RAR}}(\phi_0)|$ and $|T_{\text{vb}}(\phi_0)|$ are located in the areas on the right and left of the point $\phi_0 = 90°$, i.e. closer to $\phi_0 = 90°$ than to $\phi_0 = 0°, 180°$. Obviously, the same takes place in the interval $180° < \phi_0 < 360°$.

In Figure 6.4 (reproduced from Brüning et al. (1990)), where the calculations were applied to SEASAT SAR, we see the phases of linear SAR MTS in two parts; one of these phases, namely, η_{RAR} is in the form

$$\eta_{\text{RAR}} = \eta_{\text{RAR}}^{\text{max}} \sin \phi_0 \qquad (6.83)$$

where $\eta_{\text{RAR}}^{\text{max}} = 45°$. Such value of $\eta_{\text{RAR}}^{\text{max}}$ complies with the supposition that the intensity maximum of resonance ripples is located directly at the top of a large wave ($\eta_{\text{hydr}} = 0$)

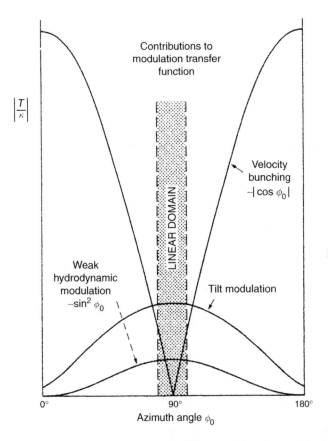

Figure 6.3 The dependence of the non-dimensional MTFs describing tilt, weak hydrodynamic and velocity bunching modulation on azimuthal angle Φ_0. There is a small angular interval around the range direction (in vicinity $\Phi_0 = 90°$), where velocity bunching is a linear mapping process (Alpers et al. 1981).

besides $|T_{\text{hydr}}| = |T_{\text{tilt}}|$. If we consider a value closer to experimental data $\eta_{\text{hydr}} = 30°$, then we have $\eta_{\text{RAR}}^{\text{max}} = 60°$. In this case in the areas adjoining the points $\phi_0 = 90°$ and $\phi_0 = 270°$, on the left and right, i.e. in the maxima of $A(\phi_0)$, we get $|\Delta\eta| \approx 90 \pm 10°$. Consequently, in the maximum positions of $A(\phi_0)$ the value of $\cos\Delta\eta$ does not exceed 0.2 and therefore within the entire interval $0° \leq \phi_0 \leq 360°$ we obtain

$$\left|T_{\text{SAR}}^{\text{lin}}(\vec{\kappa})\right|^2 \approx \left|T_{\text{RAR}}(\vec{\kappa})\right|^2 + \left|T_{\text{vb}}(\vec{\kappa})\right|^2 \tag{6.84}$$

In accordance with these estimates, the SAR image spectrum itself can be written as

$$\hat{\tilde{W}}_{\text{i,SAR}}^{\text{lin}}(\vec{\kappa}) = \hat{\tilde{W}}_{\text{RAR}}(\vec{\kappa}) + \langle\hat{\sigma}_0\rangle^2 \left(\frac{R}{V}\right)^2 \kappa_y^2 \hat{\tilde{W}}_{\text{rad}}(\vec{\kappa}) \tag{6.85}$$

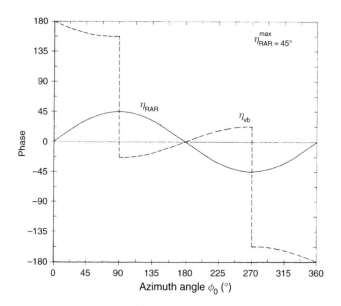

Figure 6.4 Phases η_{RAR} of the RAR MTF and η_{vb} of the velocity bunching MTF as a function of the azimuthal angle ϕ_0 (Brüning et al. 1990).

Having taken into account that

$$\hat{W}_{rad}(\vec{\kappa}) = \frac{1}{(2\pi)^2} \int d\vec{\rho} B_{rad}(\vec{\rho}) \exp(-i\vec{\kappa}\vec{\rho}) \qquad (6.86)$$

and switching over from the correlation function $B_{rad}(\vec{\rho})$ to the correlation coefficient $\hat{r}_{rad}(\vec{\rho})$, we obtain

$$\hat{W}^{lin}_{i,SAR}(\vec{\kappa}) = \hat{W}_{RAR}(\vec{\kappa}) + \langle\hat{\sigma}_0\rangle^2 (2\pi)^{-2} \left(\frac{R}{V}\right)^2 \hat{\sigma}^2_{rad} \kappa^2_y \int d\vec{\rho} \exp(-i\vec{\kappa}\vec{\rho}) \hat{r}_{rad}(\vec{\rho}) \qquad (6.87)$$

Naturally, a question arises as to the applicability area of the linear approximation. One can easily see that conditions (6.71) and (6.73), which had to be satisfied for the linear approximation (6.75), match together in essence, though Eqn (6.73) as opposed to (6.71) is an accurate criterion. Sometimes these two inequations are interpreted as a smallness condition of SAR imagery average shift against the surface characteristic wavelength (in the azimuthal projection). This condition is believed to be accomplished if the image shift is not more than a fourth of the wavelength (Hasselmann et al. 1985).

However, after a more detailed look we see that condition (6.73) of SAR imagery linearity is not as mild as it seems at first sight.

Let us examine a monochromatic wave propagating azimuthally. Then

$$\left|\frac{d\widehat{v}_{\text{rad}}}{dy}\right|_{\text{max}} = \frac{2\pi}{\Lambda_{v}}\left|\widehat{v}_{\text{rad}}\right|_{\text{max}} \approx \frac{4\pi}{\Lambda_{v}}\widehat{\sigma}_{\text{rad}} \qquad (6.88)$$

and Eqn. (6.73) becomes

$$\frac{R}{V}\frac{\widehat{\sigma}_{\text{rad}}}{\Lambda_{v}} \leq \frac{0.3}{4\pi} \qquad (6.89)$$

i.e. the shift has to be less than 0.025Λ. If the waves are not monochromatic, condition (6.73) grows still more stringent, as Λ_{v}, i.e. the characteristic wavelength in the orbital velocity spectrum can turn out twice or three times less than the surface characteristic wavelength Λ_{0} depending on the roughness type (see below).

Notably, for ERS-1 and ERS-2 European satellites $R/V \approx 120c$ and, consequently, at $\widehat{\sigma}_{\text{rad}} \approx 0.5\text{ms}^{-1}$ and $\Lambda_{0} \approx 100\text{m}$, the linearity condition is fulfilled only for the roughness travelling very close along or against the X-axis, when the imaging is done by backscatter cross-section fluctuations, and not by orbital velocities. If the wave travelling direction considerably differs from the range one, linear approximation works only for slightly sloping swell.

West et al. (1990) also came to the conclusion based on the numerical modelling data that SAR imaging process is highly non-linear under most realistic conditions. They state that the surface wave spectral information cannot be extracted from the image using a linear transfer function.

Non-linear nature of SAR imaging mechanisms can also, to some degree, taken into account with the so-called quasi-linear approximation (Hasselmann and Hasselmann 1991).

Comparing expressions (6.34) and (6.69) and omitting insignificant multipliers before integrals, we see that these formulae are identical in their structure. The difference is only that Eqn. (6.34) has unsmoothed values and nominal SAR resolution cell, while in Eqn. (6.69) there are the smoothed ones and the smeared ones. Consequently, if in Eqn. (6.49) we introduce corresponding replacements, we will obtain an expression for SAR image spectrum averaged over the sub-resolution scales:

$$\widehat{W}_{1,\text{SAR}}(\vec{\kappa}) \propto \exp\left\{-\left[\frac{\Delta_{\text{SAR}}^{2}}{2\pi^{2}} + \left(\frac{R}{V}\right)^{2}\widehat{\sigma}_{\text{rad}}^{2}\right]\kappa_{y}^{2}\right\}$$

$$\times\left\{\widehat{\widetilde{W}}_{\text{RAR}}(\vec{\kappa}) + (2\pi)^{-2}\sum_{n=1}^{\infty}\frac{1}{n!}\left(\frac{R}{V}\right)^{2n}\widehat{\sigma}_{\text{rad}}^{2n}\kappa_{y}^{2n}\int d\vec{\rho}\exp(-i\vec{\kappa}\vec{\rho})\widehat{B}_{\text{RAR}}(\vec{\rho})\widehat{r}_{\text{rad}}^{n}(\vec{\rho})\right\} \quad (6.90)$$

Taking into account

$$B_{\text{RAR}}(\vec{\rho}) = \langle\widehat{\sigma}_{0}\rangle^{2} + \widehat{b}_{\text{RAR}}(\rho) \qquad (6.91)$$

we notice that in the curly brackets the second multiplicand in expression (6.85), among other summands has the image spectrum in its linear approximation (6.87):

$$
\hat{\tilde{W}}_{\text{RAR}}(\vec{\kappa}) + (2\pi)^{-2} \sum_{n=1}^{\infty} \frac{1}{n!} \left(\frac{R}{V} \right)^{2n} \hat{\sigma}_{\text{rad}}^{2n} \kappa_y^{2n} \int \mathrm{d}\,\vec{\rho} \exp(-i\vec{\kappa}\,\vec{\rho}) \hat{B}_{\text{RAR}}(\vec{\rho}) \hat{r}_{\text{rad}}^n(\vec{\rho})
$$

$$
= \hat{\tilde{W}}_{\text{i,SAR}}^{\text{lin}}(\vec{\kappa}) + (2\pi)^{-2} \sum_{n=2}^{\infty} \frac{1}{n!} \left(\frac{R}{V} \right)^{2n} \hat{\sigma}_{\text{rad}}^{2n} \kappa_y^{2n} \int \mathrm{d}\,\vec{\rho} \exp(-i\,\vec{\kappa}\,\vec{\rho}) \hat{r}_{\text{rad}}^n(\vec{\rho})
$$

$$
+ (2\pi)^{-2} \sum_{n=1}^{\infty} \frac{1}{n!} \left(\frac{R}{V} \right)^{2n} \hat{\sigma}_{\text{rad}}^{2n} \kappa_y^{2n} \int \mathrm{d}\,\vec{\rho} \exp(-i\,\vec{\kappa}\,\vec{\rho}) \hat{b}_{\text{RAR}}(\vec{\rho}) \hat{r}_{\text{rad}}^n(\vec{\rho}) \qquad (6.92)
$$

After we ignore all the summands except the first one on the right-hand side of Eqn. (6.82), we obtain SAR image spectrum in quasi-linear approximation (Hasselmann and Hasselmann 1991):

$$
\hat{\tilde{W}}_{\text{I,SAR}}^{\text{q.lin}}(\vec{\kappa}) = \hat{\tilde{W}}_{\text{i,SAR}}^{\text{lin}}(\vec{\kappa}) \exp \left\{ - \left[\frac{\Delta_{\text{SAR}}^2}{2\pi^2} + \left(\frac{R}{V} \right)^2 \hat{\sigma}_{\text{rad}}^2 \right] \kappa_y^2 \right\} \qquad (6.93)
$$

Unfortunately, we do not know the exact applicability area limits of quasi-linear approximation. However, Hasselmann and Hasselmann (1991) state (quoting a private source) that the quasi-linear approximation works in the most real situations.

6.5.3 Non-linear velocity bunching mechanism

Recall that Eqn. (6.75) refers to the case when expression (6.70) has only one solution. The number of these solutions is a critical issue, and we will scrutinize it.

We introduce the parameter

$$
\beta_v = \frac{R \hat{\sigma}_{\text{rad}}}{V \Lambda_v} |\cos \phi_0| \qquad (6.94)
$$

which is the ratio of the imagery average shift to the characteristic wavelength (in SAR carrier flight direction) in the spectrum of orbital velocities caused by large-scale waves. We rewrite the condition of solution unicity in Eqn. (6.71) as

$$
\beta_v \ll 1 \qquad (6.95)
$$

and estimate the value β_v in the context of monodirectional roughness propagating azimuthally, i.e. along or opposite to the SAR platform flight direction. In this case there are no backscatter cross-section fluctuations, and the surface is only imaged via orbital velocities.

Let roughness be described by the Pierson–Moskovitz spectrum,

$$
W_z(\dot{\kappa}, \phi) = S_z(\dot{\kappa}) \delta(\phi - \phi_0) \qquad (6.96)
$$

$$S_z(\dot{\kappa}) = (\alpha/2)\kappa_0^{-2}\dot{\kappa}^{-3}\exp(-1.25\dot{\kappa}^{-2}) \tag{6.97}$$

Here $S_z(\dot{\kappa})$ is the omnidirectional spectrum (see Chapter 2), $\alpha = 8.1 \times 10^{-3}$, $\dot{\kappa} = \kappa/\kappa_0$, $\kappa_0 = \Omega_0^2/g$, Ω_0 is the frequency corresponding to the roughness temporal spectrum maximum, $g = 9.8\,\mathrm{m\,s^{-2}}$ and the angle ϕ_0 is held zero. The spectrum of large-scale orbital velocity is given by

$$W_v(\dot{\kappa}) = S_v(\dot{\kappa})\exp\left(-\frac{\dot{\kappa}^2}{\dot{\kappa}_{\mathrm{PCA}}^2}\right)\delta(\phi - \phi_0) \tag{6.98}$$

$$S_v(\dot{\kappa}) = g\kappa_0\dot{\kappa}S_z(\dot{\kappa})$$

where $\dot{\kappa}_{\mathrm{SAR}} = \Lambda_0/2\Delta y_0$ and $\Lambda_0 = 2\pi/\kappa_0$.

The value Λ_v, which is the characteristic wavelength in the orbital velocity spectrum, is defined by

$$\Lambda_v = \frac{2\pi}{\langle\kappa^2\rangle^{1/2}}, \quad \langle\kappa^2\rangle = \frac{1}{\sigma_v^2}\int_0^\infty d\kappa\,\kappa^2 S_v(\kappa) \tag{6.99}$$

where

$$\sigma_v^2 = \int_0^\infty d\kappa\,S_v(\kappa) \tag{6.100}$$

It yields

$$\beta_v = \frac{1}{2\pi}\frac{R}{V}\left[\int_0^\infty d\kappa\,\kappa^2 S_v(\kappa)\right]^{1/2}\cos\theta_0 \tag{6.101}$$

With the help of the Pierson–Moskovitz spectrum and the relation

$$\Omega_0 = 0.83\frac{g}{U} \tag{6.102}$$

where U is the near-surface wind speed, and using also the dispersion correlation $\Omega_0^2 = 2\pi g/\Lambda_0$, we obtain

$$\beta_v \approx 7.8 \times 10^{-3}\dot{\kappa}_{\mathrm{SAR}}^{1/2}\exp\left(-\frac{\sqrt{1,25}}{\dot{\kappa}_{\mathrm{SAR}}}\right)\frac{g}{U}\frac{R}{V}\cos\theta_0 \tag{6.103}$$

The parameter $\dot{\kappa}_{\mathrm{SAR}} = \Lambda_0/2\Delta_{0,\mathrm{SAR}}$ depends on the wind speed through Λ_0. According to Eqn. (6.102), $\Lambda_0 \approx 9.3U^2/g$, and therefore

$$\beta_v \approx 1.7 \times 10^{-2}\exp\left(-0,24\frac{g\Delta_{0,\mathrm{SAR}}}{U^2}\right)\left(\frac{g}{\Delta_{0,\mathrm{SAR}}}\right)^{1/2}\frac{R}{V}\cos\theta_0 \tag{6.104}$$

The dependence of β_v on the wind speed was computed at $R/V = 120c$, $\theta_0 = 30°$, $\Delta_{0,\,\mathrm{SAR}} = 7.5$ and $30\,\mathrm{m}$. These parameter values are typical of space-borne SAR. Remember that the nominal value $7.5\,\mathrm{m}$ is taken here as maximum, although for aperture synthesis the whole main lobe of the radar antenna pattern is deployed, and $30\,\mathrm{m}$ correspond to the quarter of the main lobe. The second mode is used to level down speckle noise by incoherently summing several images of one and the same surface patch that have been received from different parts of the main lobe.

Computation results from Eqn. (6.104) are shown in Figure 6.5. We can see that in the first case, condition (6.95) at significant wind is certainly not to be satisfied, and in the second case the curve is below the level $\beta_v = 1$.

However Eqn. (6.95) is not a rigid criterion; therefore, to make more definite conclusions about the applicability area of expression (6.72) we refer to Kanevsky (1993), where assuming the value $\tilde{v}_{\mathrm{rad}}$ is normally distributed, the author found the average number $\langle N \rangle$ of the equation roots in Eqn. (6.70) as a function β_v (see Appendix B):

$$\langle N \rangle = 2\sqrt{2\pi}\beta_v \exp\left[-\left(2\sqrt{2\pi}\beta_v\right)^{-2}\right] + \mathrm{erf}\left[\left(2\sqrt{2\pi}\beta_v\right)^{-1}\right] \tag{6.105}$$

where

$$\mathrm{erf}(x) = \frac{2}{\sqrt{\pi}} \int\limits_0^x e^{-t^2}\,\mathrm{d}t \tag{6.106}$$

This dependence is shown in Figure 6.6. If we take as a criterion at which $N > 1$ can scarcely take place, for instance, the condition $\langle N \rangle \leq 1.5$, i.e. $\beta_v \leq 0.3$, we can conclude from Figure 4.3 that at $\Delta_{0,\,\mathrm{SAR}} = 30\,\mathrm{m}$, Eqn. (6.75) works within $U \leq 7.5\,\mathrm{m\,s}^{-1}$. However

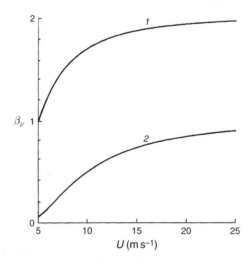

Figure 6.5 Dependence of the parameter β_v on wind speed for $\Delta_{0,\,\mathrm{SAR}} = 7.5\,\mathrm{m\,s}^{-1}$ (curve 1) and $\Delta_{0,\,\mathrm{SAR}} = 30\,\mathrm{m}$ (curve 2).

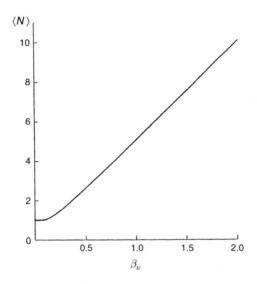

Figure 6.6 Dependence of the mean number $\langle N \rangle$ of roots of Eqn. (6.70) on the parameter β_v.

in the context of obtaining the roughness spectrum, we can hardly find the resolution $\Delta_{0,\text{SAR}} = 30\,\text{m}$ sufficient at low wind, i.e. at $\Lambda_0 \leq 50\,\text{m}$. Besides, as given in Alpers and Brüning (1986), the real resolution cell turns out a lot larger than the nominal one due to the sub-resolution orbital velocities.

Thus we found out the number of roots of Eqn. (6.70) is as a rule over a unity. For this reason provided $\pi^{-1}\Delta_{\text{SAR}} \ll \Lambda_v$ the expression for SAR signal intensity in the given case of wind waves travelling along the Y-axis is given by

$$I_{\text{i}} \propto \hat{\sigma}_0 \sum_{n=1}^{N} \left| 1 + \frac{R}{V} \frac{\mathrm{d}\hat{v}_{\text{rad}}}{\mathrm{d}y} (y_n) \right|^{-1} \tag{6.107}$$

where y_n is the nth root of Eqn. (6.70), or, at the arbitrary direction of wave propagation,

$$I_{\text{i}} \propto \sum_{n=1}^{N} \hat{\sigma}_0(y_n) \left| 1 + \frac{R}{V} \frac{\partial \hat{v}_{\text{rad}}}{\partial y} (y_n) \right|^{-1} \tag{6.108}$$

Thus, the set of random values whose fluctuations take part in the SAR imaging of roughness, apart from radar cross section and $\partial \hat{v}_{\text{rad}}/\partial y$, include N – a random number of roots of Eqn. (6.70). Physically, the fact that $N > 1$ means that an image point sums up the signals from several surface spots in the vicinity of points y_n, whose images shift and overlap in a random manner, i.e. the resolution cell disintegrates along the Y-axis into N sub-areas concentrated mainly in the area $Vt \pm (R/V)\hat{\sigma}_{\text{rad}}$.

This statement is illustrated by the computation results in Figures 6.7 and 6.8 obtained on the basis of roughness numerical model corresponding to the Pierson–Moskovitz spectrum (see Chapter 2.). These figures display fully developed roughness propagating azimuthally at $U = 10\,\mathrm{m\,s^{-1}}$ (Figure 6.7) and $U = 6\,\mathrm{m\,s^{-1}}$ (Figure 6.8), $R/V = 120\,\mathrm{s}$ and $\theta_0 = 30°$; they also show the intersection between the straight line $Vt - y'$ and random curve $(R/V)v_{\mathrm{rad}}$ as well as $\exp(-w^2)$, integrand of integral (6.30) at $\sigma_0 = \mathrm{const.}$; $\Delta_{0,\mathrm{SAR}} = 7.5\,\mathrm{m}$.

As we see, the number of subareas resulting from the resolution cell disintegration depends on the large-scale modulations of function $v_{\mathrm{rad}}(y')$ or, in other words, the modulations of the smoothed function $\widetilde{v}_{\mathrm{rad}}(y')$. However, each sub-area size depends not only on the derivative value $\partial\widetilde{v}_{\mathrm{rad}}/\partial y'$ in the intersection point of the straight line and the smoothed random curve, but also on the small-scale fluctuations of the orbital velocity in vicinity of the intersection point.

Note that at the normal distribution of orbital velocities the correlation below holds true (see Appendix C):

$$\left\langle \left| 1 + \frac{R}{V}\frac{\partial\widetilde{v}_{\mathrm{rad}}}{\partial y} \right| \right\rangle = \langle N \rangle \tag{6.109}$$

i.e. the size of each resolution cell subarea along the Y-axis is approximately $\Delta_{\mathrm{SAR}}/\langle N \rangle$.

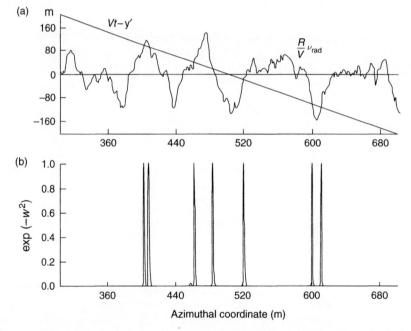

Figure 6.7 Crossings of straight line $Vt - y'$ and random function $(R/V)v_{\mathrm{rad}}$ for $(R/V) = 120\,\mathrm{s}$ and $\theta_0 = 30°$; wind roughness at the wind speed $U = 10\,\mathrm{m\,s^{-1}}$ (a). SAR resolution cell function $\exp[-w^2(y')]$ for $\Delta_{0,\mathrm{SAR}} = 7.5\,\mathrm{m}$ (b).

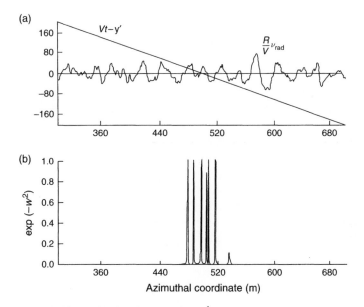

Figure 6.8 Same as in Figure 6.7, but for $U = 6\,\mathrm{m\,s^{-1}}$.

It is now clear that there are three imaging mechanisms associated with the fluctuations of various random values:

1. Backscatter cross-section fluctuations caused by tilt and hydrodynamic modulation – it is a linear mechanism common for both SAR and RAR.
2. The fluctuations of the value $\partial\widehat{v}_{\mathrm{rad}}/\partial y$ or those of the scatterers' effective density. This velocity bunching mechanism, although linear at rather low β_v, becomes non-linear with the growth of β_v.
3. The fluctuations of N – this velocity bunching mechanism is in principle non-linear, as these fluctuations are the result of the orbital velocity spectrum non-linear transformation.

As N, unlike $\partial\widehat{v}_{\mathrm{rad}}/\partial y$, is a non-local function of surface coordinates, the statistical properties of these two random values are entirely different.

Let us consider the properties of the pure nonlinear imaging mechanism associated with the fluctuations of N (Kanevsky 1993). According to the theory of non-linear transformations of random processes (Tikhonov 1986) the correlation function

$$B_N(V\tau) = \langle N(x,y)N(x,y+V\tau)\rangle \tag{6.110}$$

can be written as

$$B_N(V\tau) = \iint \mathrm{d}y_1\mathrm{d}y_2 \iint \mathrm{d}\eta_1\mathrm{d}\eta_2|\eta_1||\eta_2|p\left[\frac{V}{R}(y-y_1), \frac{V}{R}((y+V\tau)-y_2); \eta_1 - \frac{V}{R}, \eta_2 - \frac{V}{R}\right] \tag{6.111}$$

Here

$$p[\cdot] = p\left[\hat{v}_{\text{rad}}(x, y_1), \hat{v}_{\text{rad}}(x, y_2); \frac{\partial \hat{v}_{\text{rad}}}{\partial y}(x, y_1), \frac{\partial \hat{v}_{\text{rad}}}{\partial y}(x, y_2)\right] \qquad (6.112)$$

denotes the joint PDF of the process $\hat{v}_{\text{rad}}$ and its derivative at points (x, y_1) and (x, y_2); $\eta_{1,2} = (\partial \hat{v}_{\text{rad}}/\partial y)(x, y_{1,2}) + (V/R)$.

Formula (6.111) has been used for the calculation of the nonlinear fluctuations of N for the monodirectional waves propagating in azimuthal direction. The computations have been carried out for the Gaussian joint PDF whose formula contains the correlation function of $\hat{v}_{\text{rad}}(y)$. This correlation function was obtained as a Fourier transformation of the JONSWAP-type spectrum truncated by introducing the multiplier $\exp(-\kappa^2/\kappa_{\text{SAR}}^2)$. According to Alpers (1983) containing the numerical modelling of SAR signal backscattered from the ocean surface, for our calculations we considered the following values of the enhanced factor (see Eqn. (1.23)) for various roughness types:

$$\gamma = \begin{cases} 1, & \text{fully developed windsea} \\ 3.3, & \text{developing windsea} \\ 16, & \text{swell} \end{cases} \qquad (6.113)$$

Assuming $\kappa_{\text{SAR}} = 2\pi/0.1\Lambda_0$ we obtain Λ_v according to Eqn. (6.99) and then we obtain

$$\frac{\Lambda_v}{\Lambda_0} = \begin{cases} 0.3, & \text{fully developed windsea} \\ 0.33, & \text{developing windsea} \\ 0.4, & \text{swell} \end{cases} \qquad (6.114)$$

All the calculations for the numerical function $B_N(V\tau)$ carried out with the help of formula (6.111) were checked so as to fulfil the correlation

$$B_N(V\tau \to \infty) = \langle N \rangle^2 \qquad (6.115)$$

where $\langle N \rangle$ is defined by formula (6.105).

Figures 6.9–6.11 show the fluctuation spectra of the random value $N(y)$ for the three roughness types. In these figures, we considered as a parameter a more descriptive value $\beta_\zeta = R\hat{\sigma}_{\text{rad}}/V\Lambda_0$ than β_v. This value is the ratio of the average shift to the surface wave characteristic length, although, as mentioned above clearly shows, the key role belongs exactly to the parameter β_v, since SAR inherent imaging characteristics are defined by the orbital velocity field rather than the surface elevations.

The spectrum (6.98) gives the following expression for RMS of the radial component of the orbital velocity averaged (smoothed) over sub-resolution scale:

$$\hat{\sigma}_{\text{rad}} = 6.8 \times 10^{-2} U \exp\left(-0.24 \frac{g\Lambda_{0,\text{SAR}}}{U^2}\right) \left(\cos^2\theta_0 + \sin^2\phi_0 \sin^2\theta_0\right)^{1/2} \qquad (6.116)$$

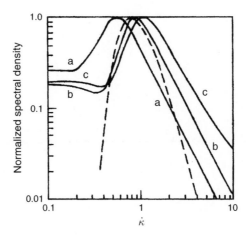

Figure 6.9 Spectrum of the random value $N(y)$ for fully developed windsea (solid lines) at $\beta_\zeta = 0.5$ (a), $\beta_\zeta = 0.25$ (b) and $\beta_\zeta = 0.17$ (c). The dotted line is the surface elevation spectrum (Kanevsky 1993).

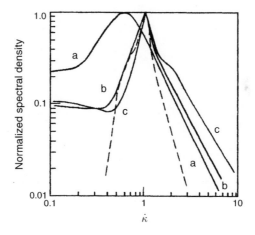

Figure 6.10 Same as in Figure 6.9 but for developing windsea.

Taking into account $\Lambda_0 = 9.3 U^2/g$ we have

$$\beta_\zeta = 0.74 \times 10^{-2} \frac{R}{V} \frac{g}{U}$$
$$\times \exp\left(-0.24 \frac{g\Delta_{0,\text{SAR}}}{U^2}\right) \left(\cos^2\theta_0 + \sin^2\phi_0 \sin^2\theta_0\right)^{1/2} |\cos\phi_0| \qquad (6.117)$$

Figure 6.12 shows the diagrams $\beta_\zeta(U)$ for two values of $\Delta_{0,\text{SAR}}$ at the azimuthal ($\phi_0 = 0°$) or anti-azimuthal ($\phi_0 = 180°$) wave propagation direction.

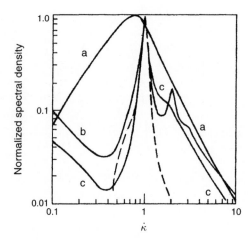

Figure 6.11 Same as in Figures 6.9 and 6.10, but for swell.

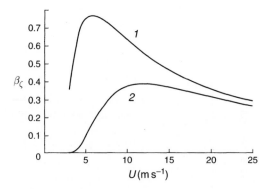

Figure 6.12 Dependence of parameter β_ζ on wind speed for two values of $\Delta_{0,\mathrm{SAR}}$ at azimuthal (or anti-azimuthal) wave travelling direction.

The analysis of the curves in Figures 6.9–6.11 indicates that the investigated mechanism highlights the major particularities of the SAR image revealed with numerical and field experiments.

1. The imagery spectrum is a lot "whiter" than the roughness spectrum. Particularly, in the low wave numbers area there are components missing in the initial roughness spectrum. This can present certain difficulties when interpreting the ocean surface imagery.
2. As the parameter β_ζ increases, the main peak of the image spectrum shifts first to the high wave number and then to the low wave number side. This effect is more pronounced, the wider the roughness spectrum. The peak shifts to the low wave number side at $\beta_\zeta \geq 0.2$, which, as Figure 6.12 indicates, at $R/V = 120c, \theta_0 = 30^\circ, \Delta_{0,\mathrm{SAR}} = 30\,\mathrm{m}$ (space SAR) corresponds to the broad range of wind speeds: $6.5\,\mathrm{m\,s^{-1}} < U < 35\,\mathrm{m\,s^{-1}}$;

therefore, the low wave number shift is to prevail over the high wave number one. At a non-azimutal wave propagation direction an apparent curtailment of the azimuthal component of the image spectrum peak radius vector makes it (radius vector) rotate towards ground range direction (i.e. towards the X-axis) – this effect as a rule is common of ocean surface imagery taken from outer space (see Figure 6.1). Non-monotonous dependence of spectrum peak shift on β_ζ, proved also in the course of numerical simulation (Brüning et al. 1990), is due to the influence (mentioned in Section 6.4) of the two competing factors on SAR image spectrum in expression (6.49).
3. The second harmonic of the main wave corresponding to the roughness spectrum peak becomes apparent only at a very narrow wave spectrum and very low β_ζ (see also Alpers 1983). Therefore the absence of higher harmonics in the real spectra of SAR ocean images, the fact that has often been used as an argument for the linearity of SAR imaging mechanism, does not in fact prove its linear character.

Finally, as noted in Keyte and Macklin (1986), the high level of the signal variability of the space SAR SIR-B worked at a very low R/V ratio, namely $R/V = 35$ s, which is in good accordance with the computation data (Kanevsky 1993), which indicate that with the decrease of β_v the variability parameter $\sigma_N^2/\langle N \rangle^2$ drastically augments.

The explanation given above allows us to give the following description to the mechanism of SAR imaging of the ocean wind waves from space. Each image point sums up the signals received from the area characterized by ground range scale Δx and the azimuthal one about $2(R/V)\hat{\sigma}_{\mathrm{rad}}$, within which there are N "brightest" patches having the average size of $\Delta x \times \Delta_{\mathrm{SAR}}/\langle N \rangle$. The fluctuations of SAR signal intensity, proportional to the random number N, form the surface image. Clearly we mean that N is essentially greater than unity, which is typical for wind waves propagating in a direction not too close to the X-axis.

As follows from the above calculations, the cut-off phenomenon is associated exactly with this mechanism, and the azimuthal size $2(R/V)\hat{\sigma}_{\mathrm{rad}}$ of the area responsible for SAR signal corresponds to the upper filtering level of low-pass filter described by the cut-off factor (6.66).

We introduce a conditional notion of a "non-linear SAR resolution cell":

$$\Delta_{\mathrm{SAR}}^{\mathrm{nl}} = 2\frac{R}{V}\hat{\sigma}_{\mathrm{rad}} \tag{6.118}$$

Of course, unlike the linear theory, $\Delta_{\mathrm{SAR}}^{\mathrm{nl}}$ is not the SAR impulse response function. In the given case it is just the azimuthal size of a surface area responsible for the formation of SAR signal, dependent (size) on the sea state. As easily noticed, provided $\Delta_{\mathrm{SAR}}^{\mathrm{nl}} \gg \Delta_{\mathrm{SAR}}$, the following correlation takes place:

$$\kappa_{\mathrm{cut\text{-}off}}\Delta_{\mathrm{SAR}}^{\mathrm{nl}} = \pi \tag{6.119}$$

where $\kappa_{\mathrm{cut\text{-}off}}$ is the cut-off wave number in the SAR image spectrum. This correlation can be treated as a mathematical representation of the fact that the cutoff factor is nothing but the manifestation of the analysed purely non-linear mechanism in the spectral domain.

Provided

$$\Delta_{\mathrm{SAR}}^{\mathrm{nl}} \ll \Delta_{\mathrm{SAR}} \qquad (6.120)$$

superimposing of partial images is absent, i.e. the considered fully non-linear imaging mechanism ceases work. In this case the SAR signal intensity is described by Eqn. (6.72), i.e. the mechanism of fluctuations of the effective density of scatterers works.

In other words, windsea SAR image is formed mainly by the fluctuations of surface elements random number, whose images shift and overlap due to the orbital velocities. This mechanism is essentially non-linear, hence the concept of a resolution cell loses the fundamental meaning it has in the linearity theory; we can only talk about an area (in the general case non-singly connected) responsible for SAR signal. As for the effective density fluctuations of the scatterers, the role this mechanism plays in SAR imaging of steep wind waves is evidently less significant.

The comparative assessment of the influence of two non-linear imaging mechanisms can be obtained with the help of Figure 6.13a and b, which display the modelling results received with the Pierson–Moskovitz spectrum for azimuthally travelling waves at two different wind speeds. Each of these figures presents the fluctuations of intensity $\Delta I_{\mathrm{i,SAR}} = I_{\mathrm{i,SAR}} - \langle I_{\mathrm{i,SAR}} \rangle$, normalized by the average value $\langle I_{\mathrm{i,SAR}} \rangle$, and the fluctuations $\Delta N = N - \langle N \rangle$ of the intersection point number of the straight line $y - y'$ and the random curve $(R/V)v_{\mathrm{rad}}(y')$, normalized by $\langle N \rangle$. The computations were carried out at the following parameter values: $\theta_0 = 30°$, $R/V = 120\,\mathrm{s}$ and $\Delta_{0,\mathrm{SAR}} = 7.5\,\mathrm{m}$. As we see, in both cases, the curves $\Delta N(y)/\langle N \rangle$ and $\Delta I_{\mathrm{i,SAR}}(y)/\langle I_{\mathrm{i,SAR}} \rangle$ are almost identical, but for the small-scale jumps $N(y)$ is missing in $I_{\mathrm{i,SAR}}(y)$. We can suspect that the smoothing of the curve $N(y)$ is caused by the scatterers' effective density fluctuations or, which is more precise, the gradual changes in the derivative dv_{rad}/dy'. The modelling results in Figure 6.13 once again prove the prevailing role of purely non-linear velocity bunching mechanism in SAR imaging of wind waves.

So far we have considered about windsea characterized by rather large values of the parameter β_v, which is eventually associated with the relatively steep wind waves. Though Figure 6.11 gives an example for swell, the minimum value of $\beta_\zeta = 0.17$ (i.e. $\beta_v = 0.4$) there corresponds to the relatively steep swell. For slightly sloping swell the ratio between the two velocity bunching mechanisms can significantly alter.

Proceeding from $\beta_v \leq 0.3$, we evaluate the maximum height of the swell whose presence make the fluctuations of scatterer effective density the major imaging agent. This condition (see Figure 6.5) can obviously be changed only to a more stringent one, i.e. the respective value of the swell height can only lessen.

If we assume for the orbital velocity that $v_{\mathrm{orb}} = \pi H_{\mathrm{sw}}/T_{\mathrm{sw}}$, where H_{sw} and T_{sw} are the height and period of the swell, then $\tilde{\sigma}_{\mathrm{rad}} \approx (\pi H_{\mathrm{sw}}/2T_{\mathrm{sw}})\cos\theta_0$, and from the condition $\beta_v \leq 0.3$ we obtain

$$H_{\mathrm{sw}} \leq 0.3 \left(\frac{R}{V} \right)^{-1} \frac{2s}{\pi \cos\theta_0} T_{\mathrm{sw}} \Lambda_{0,\mathrm{sw}} \qquad (6.121)$$

(a)

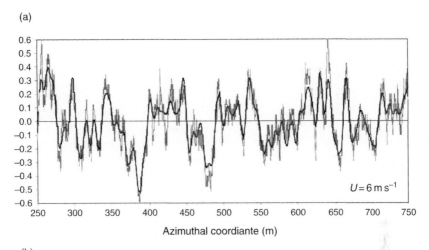

$U = 6\,\mathrm{m\,s^{-1}}$

Azimuthal coordiante (m)

(b)

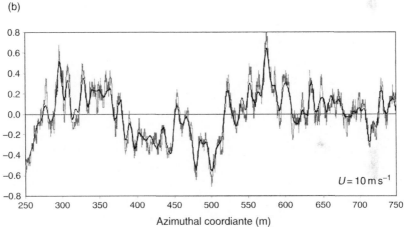

$U = 10\,\mathrm{m\,s^{-1}}$

Azimuthal coordiante (m)

Figure 6.13 Normalized fluctuations of SAR intensity (bold) and of amount of crossings for azimuthally travelling wind waves: $U = 6\,\mathrm{m\,s^{-1}}$ (a) and $U = 10\,\mathrm{m\,s^{-1}}$ (b).

where $s = \Lambda_{v,\mathrm{sw}}/\Lambda_{0,\mathrm{sw}}$ is the ratio of the characteristic wavelength in the orbital velocity spectrum to the swell wavelengths. Evidently, $s < 1$, besides, earlier (see Eqn. (6.109)) we obtained for the swell $s = 0.4$. Assuming $R/V = 120\,\mathrm{s}, \theta_0 = 30°$ and $\Lambda_{\mathrm{sw}} = 200\,\mathrm{m}$ and, consequently, $T_{\mathrm{sw}} = 11.3\,\mathrm{s}$, we get $H_{\mathrm{sw}} \leq 1.5\mathrm{m}$, i.e. the mechanism of scatterer effective density fluctuations works solely for rather flat swell.

Figure 6.14 presents modelling results analogous to those in Figures 6.7 and 6.8, yet here the modelling was applied to the swell with the wavelength of 200m and height of 1.5 m; modelling was based on the Pierson–Moskovitz spectrum. We see that in this case the area backscattering SAR signal remains singly connected and hence formula (6.72) can be used for describing the SAR image of the ocean waves.

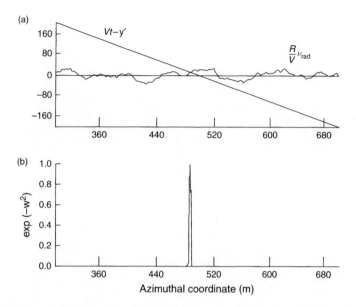

Figure 6.14 Same as in Figures 6.7 and 6.8, but for swell with the wavelength of 200 m and the height of 1.5 m.

6.5.4   SAR imaging of the mixed sea

Let us analyse the ability of SAR to "see" mixed roughness types (windsea plus swell), as commonly wind waves and swell are both present on the ocean surface. As mentioned in Chapter 2, if windsea and swell peaks are spaced rather far apart, the roughness spectrum can be seen as the sum of both component spectra.

With the help of quasi-linear approximation (applicable for a number of in situ conditions as remarked by Hasselmann and Hasselmann (1991)), we find that the image spectrum will be a sum of windsea and swell spectra transformed according to Eqn. (6.93). If the swell is rather flat at that, its quasi-linear approximation becomes linear (6.85).

However since the exact limits of quasi-linear approximation have not been defined yet, it appears quite sensible to take a broader look at the issue.

In Figure 6.15 there are (as in Figures 6.7, 6.8 and 6.14) intersection points between the straight line and the random curve for the mixed roughness case, including fully developed windsea at wind speed $U = 6\,\mathrm{m\,s}^{-1}$ and swell with the wavelength of 200 m and height of 1.5 m. Suppose that the waves are travelling azimuthally, $R/V = 120\,c$ and $\theta_0 = 30°$. As we see, the number of roots N_{mix} for mixed roughness is considerably over unity, and the swell presence brings about the quasi-periodic modulation of the number of intersection points between the straight line and the random function, so that

$$N_{\mathrm{mix}} = f(\gamma_{\mathrm{sw}})N_{\mathrm{wind}} \tag{6.122}$$

Here N_{wind} is the number of intersection points for purely wind-induced roughness and $f(\gamma_{\mathrm{sw}})$ is the modulating function dependent on the local slope γ_{sw} of the "slow" random

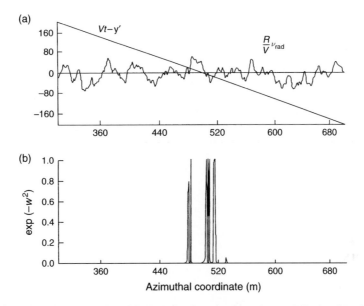

Figure 6.15 Same as in Figures 6.7, 6.8, 6.14, but for mixed roughness: fully developed windsea at wind speed $U = 6\,\mathrm{m\,s^{-1}}$ plus swell with the wavelength of 200 m and the height of 1.5 m.

curve $(R/V)v_{\mathrm{rad,sw}}(y)$, where $v_{\mathrm{rad,sw}}(y)$ is the radial component of the swell-generated orbital velocity. As there is a mechanism smoothing step change in $N(y)$, we shall hold $f(\gamma_{\mathrm{sw}})$ as a differentiable function. We expand $f(\gamma_{\mathrm{sw}})$ into a series in the vicinity of $\gamma_{\mathrm{sw}} = 0$ and retain only the linear member of the expansion; then

$$N_{\mathrm{mix}} = \left(1 + \alpha\frac{R}{V}\frac{\mathrm{d}v_{\mathrm{sw}}}{\mathrm{d}y}\right)N_{\mathrm{wind}} \qquad (6.123)$$

If the swell wavelength is significantly over the value of $2(R/V)\sigma_{\mathrm{rad,wind}}$, the coefficient α is given by the correlation

$$\langle N_{\mathrm{mix}}\rangle = \left(1 + \alpha\frac{R}{V}\frac{\mathrm{d}v_{\mathrm{sw}}}{\mathrm{d}y}\right)\langle N_{\mathrm{wind}}\rangle \qquad (6.124)$$

where $\langle N_{\mathrm{mix}}\rangle$ is the average value of N_{mix} at the fixed slope of the curve $(R/V)v_{\mathrm{rad,sw}}(y)$ in the area concentrating the intersection points (see Figure 6.14). We find the $\langle N_{\mathrm{mix}}\rangle$ value as described in Appendix B. The only difference is that here instead of $y - y'$ we should take

$$y - \left(1 + \frac{R}{V}\frac{\mathrm{d}v_{\mathrm{rad,sw}}}{\mathrm{d}y}\right)y' \qquad (6.125)$$

We easily show that if

$$\frac{R}{V}\frac{\mathrm{d}v_{\mathrm{rad.sw}}}{\mathrm{d}y} \ll 1 \qquad (6.126)$$

the correlation occurs:

$$\frac{\langle N_{\text{mix}} \rangle}{\langle N_{\text{wind}} \rangle} = 1 - \frac{R}{V} \frac{dv_{\text{rad.sw}}}{dy} \tag{6.127}$$

Hence we conclude that $\alpha = -1$.

Expression (6.123) for the spatial spectrum of the N_{mix} fluctuations yields

$$\hat{W}_{\text{mix}}(\vec{\kappa}) = \hat{W}_{\text{wind}}(\vec{\kappa}) + \left(\frac{R}{V}\right)^2 \langle N_{\text{wind}} \rangle^2 \hat{W}_{v',\text{sw}}(\vec{\kappa}) + \left(\frac{R}{V}\right)^2 \hat{W}_{v',\text{sw}} \otimes \hat{W}_{\text{wind}} \tag{6.128}$$

Here $\hat{W}_{v',\text{sw}}$ is the spectrum of the function $v_{\text{rad,sw}}(y)$ slopes, the "$\otimes$" symbol stands for the spectra convolution, and the angle brackets for the averaging process. When writing Eqn. (6.128) we paid attention to the statistical independence of swell and windsea characteristics.

Let us remember that the values v_{rad} and, consequently, $v' = dv/dy$ are in linear connection with the surface elevations; therefore, the spectrum $\hat{W}_{v',\text{sw}}$ is related to the swell elevations spectrum via the linear transfer function.

Since, as indicated above, under the windsea conditions, the fluctuations of N prevail over the fluctuations of scatterer effective density, the spectra $\hat{W}_{\text{mix}}$ and $\hat{W}_{\text{wind}}$ can be seen as SAR image spectra of respectively mixed and windsea roughness with accuracy up to the constant coefficient, i.e.

$$\hat{W}_{\text{SAR,mix}}(\vec{\kappa}) = \langle \sigma_0 \rangle^2 \left(\frac{R}{V}\right)^2 \hat{W}_{v',\text{sw}}(\vec{\kappa}) + \hat{W}_{\text{SAR,wind}}(\vec{\kappa}) + \left(\frac{R}{V}\right)^2 \hat{W}_{v',\text{sw}} \otimes \hat{W}_{\text{SAR,wind}}$$
$$\tag{6.129}$$

As we see, formula (6.129) has a more general meaning than the quasi-linear approximation. If the first term describes the swell in the linear approximation (see Eqn. (6.85)), which in case of flat long-wave swell hardly differs from the quasi-linear one, the second term is given by the windsea spectrum. For this spectrum the imaging mechanism at azimuthally travelling waves, as shown above, is always significantly non-linear; to that, it is fully non-linear and not only restricted to the cut-off factor. The third term, the convolution of swell and windsea image spectra, in the quasi-linear approximation is missing.

We evaluate the importance of the third term; to simplify the convolution operation we approximate a rather narrow even swell spectrum concentrated near $\pm \vec{\kappa}_0$, by a sum of two δ-functions. then

$$\hat{W}_{\text{SAR,mix}}(\vec{\kappa}) = \langle \sigma_0 \rangle^2 \left(\frac{R}{V}\right)^2 \hat{W}_{v',\text{sw}}(\vec{\kappa}) + \hat{W}_{\text{SAR,wind}}(\vec{\kappa})$$
$$+ \left(\frac{R}{V}\right)^2 \left\langle \left(\frac{dv_{\text{sw}}}{dy}\right)^2 \right\rangle \sum_{\mp} \hat{W}_{\text{SAR,wind}}(\vec{\kappa} \mp \vec{\kappa}_0) \tag{6.130}$$

Thus, the SAR imagery spectrum for mixed roughness consists of the curve $v_{\text{rad.sw}}(y)$ slope spectrum (with weight coefficient) and the sum of two other members on the

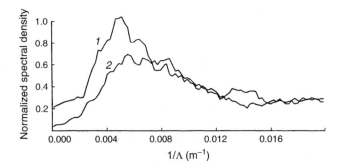

Figure 6.16 Image spectrum of mixed roughness (wind waves at the wind speed $U = 6\,\mathrm{m\,s^{-1}}$ together with the swell with the wavelength 200 m and the height 1.5 m; curve 1) we can also see here the windsea image spectrum at the wind speed $U = 6\,\mathrm{m\,s^{-1}}$ (curve 2).

right-hand side of Eqn. (6.130) perceived as a one whole windsea spectrum, a little bit broadened as compared with $\hat{W}_{\mathrm{SAR,wind}}$. For the narrow spectrum of swell the maximum of the first item is at the spatial frequency close to the swell frequency.

In Figure 6.16 we see the mixed roughness image spectrum obtained by numerical modelling (windsea at the wind speed $U = 6\,\mathrm{m\,s^{-1}}$ plus swell with the wavelength 200 m and height 1.5 m (curve 1); here, we also find the windsea image spectrum at wind speed of $U = 6\,\mathrm{m\,s^{-1}}$ (curve 2). Here, as before, modelling was applied to the roughness propagating azimuthally or anti-azimuthally.

Curve 1 can be interpreted from the point of view of formula (6.130). Indeed, the mixed roughness image spectrum has a maximum at the swell spatial frequency $1/\Lambda = 0.005\mathrm{m^{-1}}$, while on the whole spectrum (curve 1) is concentrated almost where pure windsea (curve 2) spectrum is. The latter phenomenon is quite clear for two reasons. First, the swell wave number is many times less than the windsea characteristic wave number, so that the spectrum shift described by the third term on the right-hand side of Eqn. (6.130) is rather small. Second, the convolution influence is as a matter of fact small due to the smallness of the derivative $dv_{\mathrm{rad.sw}}/dy$ in the case of a flat long-wave swell.

Obviously, in case the propagation directions of the swell and wind waves do not coincide (or are not exactly opposite), the two parts of the image spectrum matching the first and other items on the right-hand side of Eqn (6.129) or (6.130), unlike Figure 6.16 will be spaced apart on the wavevector plane $\vec{\kappa}$. This means that the swell image spectrum does not usually fuse with the windsea part of the mixed roughness spectrum; besides the latter differs little from the pure wind wave image spectrum.

6.6 SPECKLE NOISE IN THE SAR IMAGE OF THE OCEAN

It is well known that the inherent attribute of SAR image of the ocean is the speckled background caused by the coherent summing of SAR antenna fields received from elementary scatterers (namely, the gravity–capillary ripples) on the ocean surface. This background, termed speckle noise, is manifested in the image spectrum as a pedestal under the sea wave image itself. This pedestal is apparently seen in Figure 6.1. Evidently, the speckle noise impairs the interpretation of SAR imagery, and scientists investigate different methods to reduce it based on a variety of speckle-noise models. However, in the context of the theory outlined here we do not need to introduce an external model of speckle noise, because the respective part of SAR image correlation function is already present as its summand B_2 (see formula (6.13)). Recall that the third summand $B_3(\vec{\rho})$ on the right-hand side of Eqn. (6.13) is negligible and hence can be neglected when compared with the first two summands.

Consider $B_2 = B_I - B_i$, where B_I and B_i are the correlation functions of the full SAR signal and of its proper imaging part, respectively. After we write the full signal as $I = I_i n$, where n is the multiplicative noise with an average value equal to unity, we see that B_2 is nothing but the correlation function of the value $\Delta I = I - I_i = I_i(n-1)$. Consequently, B_2 is the product of the correlation function of the image proper by the noise covariance function.

Strictly speaking, the definition "speckle noise" refers to the multiplicand n in the product $I_i n$. However, we shall also apply it to the speckled underlayer $\Delta I = I_i(n-1)$, associated with this noise, since in the long run it is exactly the underlayer that can be seen as hampering the imagery interpretation.

Choosing Φ_I in Eqn. (6.31), we turn to expression (6.18) and switch over to new variables:

$$
\begin{aligned}
\vec{r}_1 - \vec{r}_2 &= \vec{\rho}', & t_1 - t_2 &= \tau' \\
\vec{r}_1 + \vec{r}_2 &= 2\vec{r}', & t_1 + t_2 &= 2t'
\end{aligned}
\tag{6.131}
$$

Here the integral over spatial coordinates is taken within the cross-hatched area shown in Figure 6.17, and the integration over the temporal variables is carried out over the inside of the rhombohedron represented in Figure 6.18. Assuming, as before, the conditions $\Delta t \ll T_0, \Delta x \ll \Lambda_0$, are fulfilled, where T_0 and Λ_0 are the characteristic period and wavelength of the large-scale roughness, we write for the integral I_2:

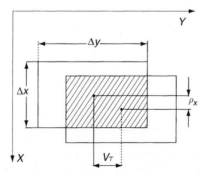

Figure 6.17 The integration area (hatched) over spatial coordinates in integral (6.18).

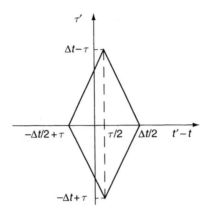

Figure 6.18 Rhombohedron interior – the integration area over temporal coordinates in integral (6.18).

$$I_2 = \Delta x \left(1 - \frac{|\rho_x|}{\Delta x}\right) \int dy' |m(x,y';t)|^2 \int_{t+\tau-\Delta t/2}^{t+\Delta t/2} dt' \exp\left(-2ik\frac{V^2}{R}\tau t'\right)$$

$$\times \int_{-\tau_0(t')}^{\tau_0(t')} d\tau' \exp\left\{-2ik\left[\frac{\partial \zeta}{\partial t}\cos\theta_0 - \frac{V}{R}\left(V\left(t+\frac{\tau}{2}\right) - y'\right)\right]\tau'\right\} \quad (6.132)$$

$$\times \int d\vec{\rho}' B_\xi(\vec{\rho}',\tau') \exp\left(-i\vec{\kappa}_{\text{res}}\,\vec{\rho}'\right)$$

$$\tau_0(t') = \begin{cases} \Delta t - 2\tau + 2(t'-t) & \left(t+\tau-\dfrac{\Delta t}{2} \le t' \le t+\dfrac{\tau}{2}\right) \\[2mm] \Delta t - 2(t'-t) & \left(t+\dfrac{\tau}{2} \le t' \le t+\dfrac{\Delta t}{2}\right) \end{cases} \quad (6.133)$$

After the transformations that are fully identical to those in Section 6.3, we get

$$I_2 = \Delta x \left(1 - \frac{|\rho_x|}{\Delta x}\right) \int dy' \sigma_0(x,y') K(x,y';t,\tau) \quad (6.134)$$

$$K = \frac{2}{b}\left\{ \int_{t+\tau-\Delta t/2}^{t+\tau/2} dt' \exp(-iat') \sin[b(\Delta t - 2\tau + 2(t'-t))] \right.$$

$$\left. + \int_{t+\tau/2}^{t+\Delta t/2} dt' \exp(-iat') \sin[b(\Delta t - 2(t'-t))] \right\} \quad (6.135)$$

Here we used the symbols:

$$a = 2k\frac{V^2}{R}\tau, \quad b = 2k\frac{V}{R}\left[V\left(t + \frac{\tau}{2}\right) - y' - \frac{R}{V}v_{\text{rad}}\right] \tag{6.136}$$

The integral K is transformed into

$$K = \frac{2}{b}\exp\left[-ia\left(t + \frac{\tau}{2}\right)\right]Re\left\{\exp\left[i\frac{a}{2}(\Delta t - \tau)\right]\int_0^{\Delta t - \tau} du\exp\left(-i\frac{a}{2}u\right)\sin bu\right\} \tag{6.137}$$

and obtain

$$K = \exp\left[-ia\left(t + \frac{\tau}{2}\right)\right]\frac{\sin w}{w}\frac{\sin w_\tau}{w_\tau} \tag{6.138}$$

$$w_\tau = \frac{\pi}{\Delta_{0.\text{SAR}}}\left[V(t + \tau) - y' - \frac{R}{V}v_{\text{rad}}\right] \tag{6.139}$$

When writing Eqns (6.138) and (6.139) we took into account that for focused SAR, i.e. for SAR with quite a long integration time, the condition $V\Delta t \gg \Delta_{0.\text{SAR}}$ holds true. Then I_2 is given by

$$I_2 = \exp\left[-2ik\frac{V^2}{R}\tau\left(t + \frac{\tau}{2}\right)\right] \times I_s \tag{6.140}$$

$$I_s = \Delta x\left(1 - \frac{|\rho_x|}{\Delta x}\right)\int dy'\sigma_0\frac{\sin w}{w}\frac{\sin w_\tau}{w_\tau} \tag{6.141}$$

As Eqn. (6.15) states, it is I_s ("s" stands for "speckle") that defines the summand B_2 of correlation function (6.13), which we rename as B_s:

$$B_s = \langle I_s^2(\vec{r}, \vec{\rho})\rangle \tag{6.142}$$

To sum up the results we set forth the major formulae of the theory:

$$B_I(\vec{\rho}) = B_i(\vec{\rho}) + B_s(\vec{\rho}) \tag{6.143}$$

$$B_i(\vec{\rho}) = \langle I_i(\vec{r})I_i(\vec{r} + \vec{\rho})\rangle \tag{6.144}$$

$$B_s = \langle I_s^2(\vec{r}, \vec{\rho})\rangle \tag{6.145}$$

Depending on the form of the multiplier Φ_t (Eqn (6.31) or (6.22)), I_i and I_s are given by the expressions

$$I_i = \begin{cases} \Delta x \int dy' \sigma_0(x, y') \left[\dfrac{\sin w(x, y')}{w(x, y')} \right]^2 & \text{for (6.31)} \\ \dfrac{\pi}{2} \Delta x \int dy' \sigma_0(x, y') \exp\left[-w^2(x, y')\right] & \text{for (6.22)} \end{cases} \qquad (6.146)$$

$$I_s = \begin{cases} \Delta x \left(1 - \dfrac{|\rho_x|}{\Delta x}\right) \int dy' \sigma_0(x, y') \dfrac{\sin w(x, y')}{w(x, y')} \dfrac{\sin w_\tau(x, y')}{w_\tau(x, y')} & \text{for (6.31)} \\ \dfrac{\pi}{2} \Delta x \left(1 - \dfrac{|\rho_x|}{\Delta x}\right) \int dy' \sigma_0(x, y') \exp\left[-\dfrac{1}{2} w^2(x, y')\right] \exp\left[-\dfrac{1}{2} w_\tau^2(x, y')\right] & \text{for (6.22)} \end{cases}$$
$$(6.147)$$

$$w = \dfrac{\pi}{\Delta_{0,\text{SAR}}} \left[Vt - y' - \dfrac{R}{V} v_{\text{rad}}(x, y') \right] \qquad (6.148)$$

$$w_\tau = \dfrac{\pi}{\Delta_{0,\text{SAR}}} \left[V(t + \tau) - y' - \dfrac{R}{V} v_{\text{rad}}(x, y') \right] \qquad (6.149)$$

Equations (6.147) hold true provided $|\rho_x| < \Delta x$, and in the case of $|\rho_x| \geq \Delta x$ we evidently get $I_s = 0$.

As mentioned above, B_i describes the roughness image proper, and B_s defines the speckle matter, i.e. the speckle noise. According to Eqn. (6.147) the characteristic range and azimuthal dimensions of the speckles are approximately Δx and $\Delta_{0,\text{SAR}}$, respectively. Notably, $B_s(0) = B_i(0) > B_i(0) - \langle I_i \rangle^2$, which means that the entire (i.e. calculated over the whole spectrum) power of SAR signal intensity speckle-noise fluctuations exceeds the summary power of the fluctuations responsible for the roughness image proper.

It is to underline once more that within the framework of our theory both roughness imagery and the speckle noise associated with it are described in frameworks of the united theory; therefore we need not introduce any external speckle-noise model extrinsic to the roughness proper imaging theory.

6.7 MODIFIED SPECTRAL ESTIMATE FOR THE SAR IMAGE OF THE OCEAN

If we proceed from the correlation function to the image spectrum, the spectrum of the roughness image proper turns out to be built upon a "pedestal" made up by speckle-noise spectrum, as follows from expressions (6.143). The speckle-noise pedestal is well illustrated in Figure 6.1, where it is only 4–5 dB below the image spectrum peak. Hence, the main part of the noise power may be concentrated in the spectral area covered by the surface roughness and might shut off spectrum details. This is the reason why the investigation of ways to bring down the speckle-noise level is of such importance.

The most widespread and popular method of speckle noise suppression is the multi-look method, where they perform the incoherent summing of several images of the same surface spot surveyed during one passage of SAR carrier. Individual look images are formed consecutively by different sectors of antenna pattern, and the speckle-noise realizations stay independent of each other at the same or slightly changing realizations of large waves to the extent that the small-scale ripple realizations alter during the passage from one sector to another. Evidently, the level of noise suppression greatly depends also on the number of summed images, which is definitely finite (3–4 in space-borne SAR) to avoid a considerable impairment of the spatial resolution. A complete theoretical validation of this processing method for SAR images of the ocean is given in Ouchi and Wang (2005).

In the recent years the "cross spectrum" method has gained more acceptance (Engen and Johnson 1995). In this method as a surface spot image spectrum they take a cross spectrum (Fourier-transformed cross-correlation function) of two images obtained for this spot in different sectors of antenna pattern. Theoretically, the method efficiency is proved by the authors of the work on the basis of the model introduced by them, according to which the full (speckle noise included) SAR image of the water surface is described by the integral transformation (6.30) applied not to the radar cross section σ_0, but to the value $n\sigma_0$, where n is the white noise. In practice, this method provides effective suppression of high wave number components of the image spectrum at the same time not affecting the peaks in low wave number range. It is to note though that transformation (6.30) has been deduced solely for speckle-noise-free RAR signal, and its applicability to the noise-like signals has not been proved yet.

Below we explore the method to make SAR image spectrum free of speckle-noise pedestal derived directly from the above outlined theory (Kanevsky 2002, 2005).

We introduce the complex value $I_c = a_{SAR}^2$, term it complex intensity and write its correlation function:

$$B_c = \langle I_c(\vec{r}) I_c^*(\vec{r} + \vec{\rho}) \rangle \tag{6.150}$$

With respect to Eqn. (6.8) we obtain

$$B_c = \frac{16k^8}{(\Delta t)^4 \pi^2} \exp[4ik(R - R_+)] \iint_{\Delta t} dt_1 dt_2 \int\int_{\Delta(t+\tau)} dt_3 dt_4 \iint_{\Delta \vec{r}} d\vec{r}_1 d\vec{r}_2 \int\int_{\Delta(\vec{r}+\vec{\rho})} d\vec{r}_3 d\vec{r}_4$$

$$\times \Big\langle m(\vec{r}_1, t_1) \times m(\vec{r}_2, t_2) m^*(\vec{r}_3, t_3) m^*(\vec{r}_4, t_4) \xi(\vec{r}_1, t_1) \xi(\vec{r}_2, t_2) \xi(\vec{r}_3, t_3) \xi(\vec{r}_4, t_4)$$

$$\times \exp\{2ik[(x_1 + x_2 - x_3 - x_4 + 2\rho_x) \sin \theta_0$$

$$- [\zeta(\vec{r}_1, t_1) + \zeta(\vec{r}_2, t_2) - \zeta(\vec{r}_3, t_3) - \zeta(\vec{r}_4, t_4)] \cos \theta_0]\}$$

$$\times \exp\left\{ \frac{ik}{R} \left[y_1^2 + y_2^2 - 2Vt_1(y_1 - Vt) - 2Vt_2(y_2 - Vt) - 2V^2t^2 \right] \right\}$$

$$\times \exp\left\{ -\frac{2ik}{R_+} \left[y_3^2 + y_4^2 - 2Vt_3(y_3 - V(t + \tau)) - 2Vt_4(y_4 - V(t + \tau)) - 2V^2(t + \tau)^2 \right] \right\} \Big\rangle \tag{6.151}$$

Remember that $R_+ = R + \rho_x \sin\theta_0$ is the range of the point $\vec{r} + \vec{\rho}$ measured from the point of antenna dislocation at the time $t + \tau$. Substituting, as earlier, in the argument of the last exponent R_+ by R, we rewrite Eqn. (6.151) as

$$
B_c = \frac{16k^8}{(\Delta t)^4 \pi^2} e^{4i\frac{kV^2}{R}\tau(t+\tau/2)} \iint_{\Delta t} dt_1 dt_2 \iint_{\Delta(t+\tau)} dt_3 dt_4 \iint_{\Delta\vec{r}} d\vec{r}_1 d\vec{r}_2 \iint_{\Delta(\vec{r}+\vec{\rho})} d\vec{r}_3 d\vec{r}_4
$$

$$
\times \left\langle m(\vec{r}_1, t_1) m(\vec{r}_2, t_2) \times m^*(\vec{r}_3, t_3) m^*(\vec{r}_4, t_4) \xi(\vec{r}_1, t_1) \xi(\vec{r}_2, t_2) \xi(\vec{r}_3, t_3) \xi(\vec{r}_4, t_4) \right.
$$

$$
\left. \times \exp\left\{ 2ik \left[\begin{array}{l} (x_1 + x_2 - x_3 - x_4)\sin\theta_0 \\ -[\zeta(\vec{r}_1, t_1) + \zeta(\vec{r}_2, t_2) - \zeta(\vec{r}_3, t_3) - \zeta(\vec{r}_4, t_4)]\cos\theta_0 \\ +\frac{1}{2R} \left[\begin{array}{l} y_1^2 + y_2^2 - y_3^2 - y_4^2 - 2Vt_1(y_1 - Vt) - 2Vt_2(y_2 - Vt) \\ +2Vt_3(y_3 - V(t+\tau)) + 2Vt_4(y_4 - V(t+\tau)) \end{array} \right] \end{array} \right] \right\} \right\rangle .
$$

$$(6.152)$$

Averaging over the ripple realizations with the help of Eqn. (6.12), we switch from Eqn. (6.152) to the sum of three summands having grouped the variables by their indices as follows: 1324, 1423 and 1234. In this case the first two summands are equal, and the third one is negligible, so that

$$
B_c = 2\exp\left[4i\,\frac{kV^2}{R}\tau\left(t + \frac{\tau}{2}\right)\right]\langle J^2\rangle
$$

$$
J = \frac{4k^4}{\pi(\Delta t)^2} \int_{\Delta t} dt_1 \int_{\Delta(t+\tau)} dt_3 \int_{\Delta\vec{r}} d\vec{r}_1 \int_{\Delta(\vec{r}+\vec{\rho})} d\vec{r}_3\, m(\vec{r}_1, t_1) m^*(\vec{r}_3, t_3) B_\xi(\vec{r}_1 - \vec{r}_3, t_1 - t_3)
$$

$$
\times \exp\left\{ 2ik \left[\begin{array}{l} (x_1 - x_3)\sin\theta_0 - \left[\zeta(\vec{r}_1, t_1) - \zeta(\vec{r}_3, t_3)\right]\cos\theta_0 \\ +\frac{1}{2R}\left[y_1^2 - y_3^2 - 2V(t_1 y_1 - t_3 y_3) + 2V^2 t(t_1 - t_3) - 2V^2 t_3 \tau\right] \end{array} \right] \right\}
$$

$$(6.153)$$

Going back to Eqn. (6.16), we remark that $J = I_2$, and therefore taking into account Eqn. (6.140)

$$
B_c = 2e^{4i\frac{kV^2}{R}\tau\left(t+\frac{\tau}{2}\right)}\langle I_2^2(\vec{r}, \vec{\rho})\rangle = 2\langle I_s^2(\vec{r}, \vec{\rho})\rangle = 2B_s
$$

$$(6.154)$$

Thus, the noise part B_s of the image full correlation function is nothing but one-half of the complex intensity correlation function, which yields

$$
B_i(\vec{\rho}) = B_I(\vec{\rho}) - \frac{1}{2}B_c(\vec{\rho})
$$

$$(6.155)$$

Performing Fourier transformation of both the sides of Eqn. (6.155), we obtain

$$\hat{W}_i(\vec{\kappa}) = \hat{W}_I(\vec{\kappa}) - \frac{1}{2}\hat{W}_c(\vec{\kappa}) \tag{6.156}$$

Each spectrum $\hat{W}$ corresponds to its own correlation function in Eqn. (6.155).

As is known, the complex field spectrum, which corresponds to the complex intensity $I_c(\vec{r}) = Re^2 a_{SAR}(\vec{r}) - Im^2 a_{SAR}(\vec{r}) + 2iRe\, a_{SAR}(\vec{r})Im\, a_{SAR}(\vec{r})$, is a real value (Rytov et al. 1989) and

$$\hat{W}_c = \hat{W}_{c1} + \hat{W}_{c2} \tag{6.157}$$

where $\hat{W}_{c1}$ and $\hat{W}_{c2}$ are the spectra of $I_{c1} = Re^2 a_{SAR}(\vec{r}) - Im^2 a_{SAR}(\vec{r})$ and $I_{c2} = 2Re\, a_{SAR}(\vec{r})Im\, a_{SAR}(\vec{r})$, respectively.

One of the commonly accepted estimates of the spectrum is a periodogram (Marple 1987), which is usually calculated with fast Fourier transformation

$$\overline{W}_i(\vec{\kappa}) = \text{FFT}\left[I(\vec{r})\right] \tag{6.158}$$

where $I(\vec{r})$ is the signal realization (FFT is the sum of squares of Fourier sine and cosine transformations). Then Eqn. (6.156) gives the expression for the spectrum estimate free of speckle noise:

$$\hat{\overline{W}}_i(\vec{\kappa}) = \text{FFT}\left[Re^2 a_{PCA}(\vec{r}) + Im^2 a_{PCA}(\vec{r})\right]$$
$$- \frac{1}{2}\text{FFT}\left[Re^2 a_{PCA}(\vec{r}) - Im^2 a_{PCA}(\vec{r})\right] - 2\text{FFT}\left[Re a_{PCA}(\vec{r})Im\, a_{PCA}(\vec{r})\right] \tag{6.159}$$

The first term on the right-hand side of Eqn. (6.159) is a usual spectral estimate containing speckle noise, and the other two are for removing the speckle-noise pedestal.

Formula (6.155), which is the basis for estimate (6.159), holds for statistically average values, which are in fact correlation functions and their respective spectra; while estimate (6.159) is a random value subject to statistical fluctuations. As the right-hand side of Eqn. (6.155) and, consequently, the right-hand side of Eqn. (6.156) are the differences, the random value $\overline{W}_i(\vec{\kappa})$ at some points may be negative, i.e. non-physical. To get rid of negative values and obtain a steady estimate of the spectrum, the periodogram should be smoothed by any of the deployed means. There are many ways to obtain steady spectral estimates; respective information is presented in Marple (1987).

The spectral estimation method examined provides the spectrum of the ocean SAR image statistically completely free of speckle noise all over the spectrum; moreover, there is no loss in resolution.

Figure 6.19 exemplifies the application of this spectral estimation method. The experiment involved SAR imagery acquired by the ERS-2 European satellite in the $10 \times 10\,\text{km}^2$ region along the Atlantic coast of USA.

The spectrum in Figure 6.19a was obtained on the basis of a conventional spectral estimate (the first term on the right-hand side of Eqn. (6.159)) by smoothing, i.e. averaging over the matrix 5×5 points. The halo effect in the figure is in fact the speckle-noise

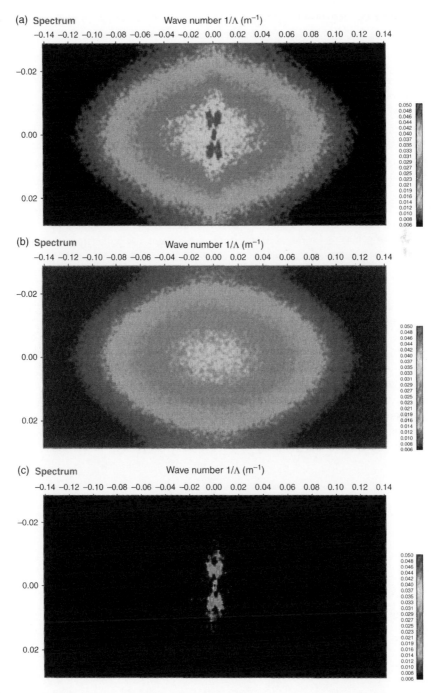

Figure 6.19 Spectrum of the SAR image (SAR ERS-2) of the ocean surface obtained by the standard method (a). Noise pedestal (b). SAR image spectrum "cleaned" from the noise pedestal (c). Asymmetry of the speckle noise pedestal is caused by asymmetry of the nominal SAR resolution cell in case of full azimuthal resolution (one-look regime).

pedestal under the roughness image spectrum proper. As the theory suggests, this pedestal is described by the sum of two last summands on the right-hand side of Eqn. (6.159); this sum is presented in Figure 6.19b. Note that the asymmetry of the speckle-noise pedestal is caused by asymmetry of the nominal SAR resolution cell in case of full azimuthal resolution (one-look regime). The speckle-noise-free spectrum is shown in Figure 6.19c (the spectra displayed in Figures 6.19b and c are also smoothed by the above-mentioned means).

In Figure 6.20 we see the cross sections of two spectra – standard and speckle-noise-free ones (each curve is normalized by its own maximum). As we see, the new method of spectral estimation gave a more pronounced spectral maximum, whose shape is drastically different from that in the speckle-noise spectrum.

Remember that the imaging theory and spectral estimation method described in this chapter are based on the resonance theory of microwave scattering by the water surface. The scattering theory (see Chapter 3) works for moderate waves and a particular range of probing angles, and should any of its application conditions be violated, all the conclusions and results lose their ground. Besides, the method can prove inefficient in the case of very faint scattering, i.e. at a fairly high level of thermal noise originated in the SAR system itself, and not described by the useful formulae.

Notably, the noise pedestal in the SAR ocean image spectrum including both the speckle-noise part and the part caused by the thermal noise was considered in Alpers and Hasselmann (1982).

As seen in Figure 6.19, the spectrum characterized by central symmetry has 180° ambiguity in wave propagation direction that is inherent in any spatial spectrum and, in particular, a single-look SAR image spectrum. To resolve this ambiguity, several multi-look techniques have been proposed (Vachon and Raney 1991, Vachon and West 1992). These techniques use the fact that individual looks correspond to slightly offset observation times during which the relative position of the imaged wave shifts along the wave propagation direction. The same objective is pursued when using the inter-look image cross spectrum (Engen and Johnsen 1995). Phase terms present in the cross spectrum

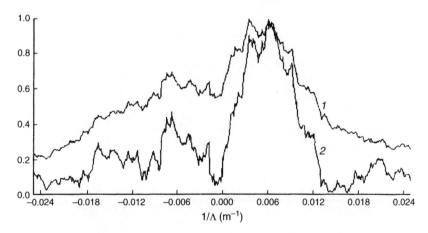

Figure 6.20 Sections of the spectra: curve 1 corresponds to the spectrum in Figure 6.19a and curve 2 to the spectrum in Figure 6.19c.

allow the wave direction to be resolved in many cases. For typical ERS or RADARSAT-1 SAR parameters, the time offset for individual looks corresponds to a few tenths of a second, which is often adequate to resolve the wave propagation direction.

The SAR image spectrum cleaned of speckle noise by any of the existing means can be in principle used to retrieve the ocean wave spectrum. There is nowadays a number of inversion algorithms intended for such retrieval (Hasselmann and Hasselmann 1991, Krogstad et al. 1994, Engen et al. 2000). Every algorithm is an iteration procedure minimizing a cost function that involves the best guess a priori ocean wave spectrum, which would usually come from a wave model, the non-linear spectral transform (quasi-linear of fully non-linear) and the observed SAR image spectrum. The same principle forms the basis of the algorithm suggested in Schulz-Stellenfleth and Lehner (2004); this algorithm unlike the previous ones works in the spatial domain, rather than spectral one. Other a priori information, such as the wind vector derived from scatterometry, can also be used in the inversion process (Mastenbroek and de Valk 2000), and the article (Lyzenga 2002) talks about an algorithm which does not use a priori information at all.

It is to highlight though that all these inversion algorithms inevitably involve the MTF, which is not just scarcely examined but also usually contains a lot of external noise to the large-scale roughness processes (see Section 3.2.2). The use of algorithms has indicated a sufficient recovery of long-wave (with more than 250 m wavelength) roughness spectrum part, which as a rule occurs in the pass band of the low-pass filter with characteristic described by cut-off factor (6.66) or (6.68).

6.8 PECULIARITIES OF THE SAR IMAGERY OF THE OCEAN SURFACE

This section touches upon particular cases of applying the studied above imaging mechanisms as specific features of SAR imagery of the ocean surface. These features are for the main part the distortions of the imaged surface structures we have to think about when interpreting SAR imagery. Besides, the last subsection is devoted to the potential of SAR as a space-borne high-resolution measuring tool for the near-surface wind speed.

6.8.1 SAR imaging of near-shore areas

This section is based mainly on the material of Wackerman and Clemente-Colon (2004). SAR with its high resolution at a fairly wide swath is extremely important for distant monitoring of the coastal regions, as these regions are characterized by a considerable spatio-temporal changeability of environmental characteristics (Johannessen 1995). A lot much has been said and written about the high potential of SAR for performing this work (see e.g. *SAR User's Manual* (2004) references therein). Yet, the specifics of SAR imaging mechanisms require a particular alertness in the task of interpreting SAR imagery of the ocean and the coastal areas in particular.

The most significant environmental feature of the near-shore region is that water depths are shallow and change rapidly. The shallow depths mean that ocean surface waves can "feel" the bottom, which causes a change in how waves move, or propagate, along the surface, and a change in the shape of the wave as the depth gets shallower (Dean and Dalrymple 1991). A wave starts to be affected by water depth when the depth is less than or equal to half the wavelength. At this point the wave starts to turn towards the shore so that eventually the wave crests become parallel to the shoreline regardless of the angle at which they started in deeper water. This turning is referred to as wave refraction, and it occurs more rapidly (i.e. over shorter lengths) for regions where the depth is getting shallow faster than in regions where the depth change is gradual.

Wave refraction can be used to estimate bathymetry, and wave propagation models exist to predict wave refraction as a function of depth (Lui et al. 1985). By estimating the wave direction and wavelength throughout the image, one can get a better estimate of water depth by inverting these models, i.e. finding a depth map that generates the observed wave directions and wavelengths. However this approach needs much thinking in the case when the source of information about the wave field is SAR image, since, as we know, apparent wave directions and wavelengths greatly depend on the distribution of surface velocities.

One of the factors perturbing the velocity field is wave breaking. In the coastal area, the wavelength and amplitude of the waves change as depth gets shallower, namely, wave height increases and wavelength decreases. Eventually, the wave gets so steep that it can no longer support the water and it breaks. Outside the coastal line the breakings of large waves take place at rather high states of the surface.

When a wave breaks, it creates a region of very rough water. The increased roughness of the water produces an increase in the amplitude of the Bragg waves within the breaking regions, and thus implies an increase of the radar cross section. Besides, the turbulence, or motion, of the water within the breaking region as well as the overall motion of the breaking region as it is being propagated along with the wave causing shifting and smearing effects in the SAR response. Although the breaking events themselves are spatially small, each of them looks in the image as a rather long bright stripe aligned in the azimuth direction (i.e. orthogonal to the SAR look direction).

Figures 6.21 and 6.22 show SAR images of one and the same off-coast segment at considerably different surface states, obtained with aircraft-based fine-resolution SAR (nominal resolution 2 m); the imaged area is approximately $500 \times 500 \, \text{m}^2$. Figure 6.21 corresponds to not very high state of the surface; here at a distance of 200–300 m off coast (in the left part of the image) we can see wave crests oriented nearly parallel to the coast line, just as the physics of the phenomena suggests (see above). As for the surf zone, here the smeared images of breaking events are oriented perpendicular to the SAR look direction. Figure 6.22 presents SAR image of the same surface spot at high wave state. In this case all the image panorama of the sea surface is covered by bright lines, each of these unlike Figure 6.21 is not a wave crest image, but the breaking event image smeared in the azimuthal direction. These lines are oriented by SAR platform flight direction, and not by the coast line. Evidently, we can scarcely draw any bathymetric information out of such pictures.

6.8.2 Peculiarities of SAR imaging of ships and slicks

A well-known particularity of SAR image of a ship and its wake (as well as a moving train and still rails) is the ship imagery displacement against its wake. From Section 6.1 we know that this displacement is in fact an azimuthally directed shift equal to $(R/V)v_{\text{ship,rad}}$

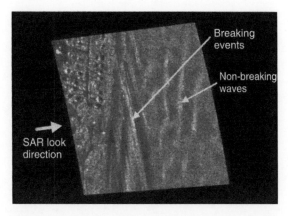

Figure 6.21 Fine resolution (2 m) airborne SAR (X-band, HH) image of breaking waves. The bright smears are signatures of breaking events. The breaking is taking place very near to the shore and non-breaking waves can be seen in the region. The imaged area is approximately $500 \times 500\,\mathrm{m}^2$ (Wackerman and Clemente-Colon 2004).

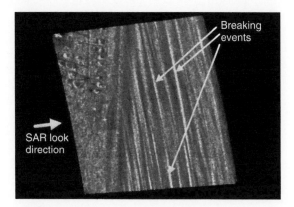

Figure 6.22 Fine resolution (2 m) airborne SAR (X-band, HH) image of breaking waves. This shows a much higher sea state than Figure 6.23. Note that the smears corresponding to breaking events occupy all imaged sea area and are aligned in the azimuthal direction (i.e. orthogonal to the SAR look direction) (Wackerman and Clemente-Colon 2004).

(along or opposite to the direction of SAR platform flight depending on the radial component sign in ship velocity $\vec{v}_{\text{ship}}$); therefore the shift value and direction allow us to obtain the value and direction of the radial component of the ship velocity. This shift is exemplified in Figure 6.23 (reproduced from Pichel et al. (2004)).

Another inherent feature that can be explained from the point of view of examining the above imaging mechanisms is the occasional disappearance of ships from SAR imagery of the ocean, although ships under the calm sea conditions are usually characterized by rather high reflectivity as compared to the water surface one.

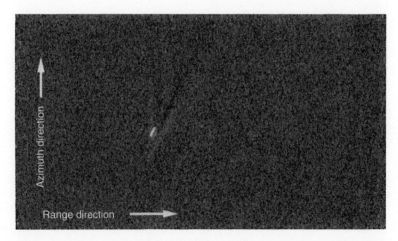

Figure 6.23 The SAR image of a ship displaced from its wake (Pichel et al. 2004). © CSA 1997.

We define the ship contrast K_{ship} as an intensity ratio of backscattered SAR signals in two cases:

$$K_{\text{ship}} = \frac{I_{\text{i}}(\text{water} + \text{ship})}{I_{\text{i}}(\text{water})} \tag{6.160}$$

The area backscattering SAR signal is composed of N "bright" sections, one of which has been replaced by the ship. Therefore

$$K_{\text{ship}} \approx \frac{I_{\text{i}}(\text{ship}) + \sigma_{0,\text{water}} S_{\text{water}}(N-1)}{\sigma_{0,\text{water}} S_{\text{water}} N} \tag{6.161}$$

where $\sigma_{0,\text{water}}$ is the NRCS of the water surface and $S_{\text{water}} \approx \Delta x \Delta_{\text{SAR}} / \langle N \rangle$ is the square of the "bright" section of the water surface. Making use of expression (6.30) and assuming that along the whole azimuthal size of the ship $L_{\text{ship}}^{\text{az}}$ the normalized cross section of the ship $\sigma_{0,\text{ship}}$ is the constant value, we obtain

$$I_{\text{i}}(\text{ship}) = \sigma_{0,\text{ship}} L_{\text{ship,il}}^{\text{range}} \Delta_{0,\text{SAR}} \frac{\sqrt{\pi}}{2} \text{erf}\left(\frac{\pi L_{\text{ship}}^{\text{az}}}{2\Delta_{0,\text{SAR}}}\right) \tag{6.162}$$

Here $L_{\text{ship,il}}^{\text{range}}$ is the ground range size of the ship's illuminated part, and $\text{erf}(x)$ is the error function:

$$\text{erf}(x) = \frac{2}{\sqrt{\pi}} \int_0^x e^{-t^2} dt \tag{6.163}$$

If we replace N by its average value, we obtain

$$K_{\text{ship}} \approx \frac{\langle N \rangle - 1}{\langle N \rangle} + K_{0,\text{ship}} \frac{\Delta_{0,\text{SAR}}}{\Delta_{\text{SAR}}} \frac{L_{\text{ship,il}}^{\text{range}}}{\Delta x} \frac{\sqrt{\pi}}{2} \text{erf} \left(\frac{\pi L_{\text{ship}}^{\text{az}}}{2\Delta_{0,\text{SAR}}} \right) \qquad (6.164)$$
$$K_{0,\text{ship}} = \frac{\sigma_{0,\text{ship}}}{\sigma_{0,\text{water}}}$$

One can see that Eqn. (6.164) contains several factors dependent on environmental conditions. At a fairly high value of $\langle N \rangle$ the first summand on the right-hand side of Eqn. (6.164) is close to unity, and if the second summand is small against a unity, then the contrast K_{ship}, expressed in decibels, is close to zero. The higher the wind speed the bigger $\sigma_{0,\text{water}}$ and, consequently, the lower $K_{0,\text{ship}}$. Finally, the wind speed growth is accompanied by the increasing smearing of SAR resolution cell. According to Eqn. (6.61c)

$$\frac{\Delta_{0,\text{SAR}}}{\Delta_{\text{SAR}}} = \left[1 + \frac{2\pi^2 (R/V)^2 \tilde{\sigma}_{\text{rad}}^2}{\Delta_{0,\text{SAR}}^2} \right]^{-1/2} \qquad (6.165)$$

where $\tilde{\sigma}_{\text{rad}}$ is RMS of sub-resolution orbital velocity radial component. The computations performed by Alpers and Brüning (1986) showed that in the case of SEASAT SAR the smeared cell can be several times the nominal one, which also brings about the lessening of $K_{0,\text{ship}}$. Hence under certain environmental conditions SAR panoramic imagery of the ocean displays only the wake of the ship instead of the vessel itself. In this case, factor such as the location of the ship hull with respect to its wake in SAR images cannot be used to estimate the ship speed; yet to measure it other parameters of its wake can be used as well (Zilman et al. 2004).

SAR imaging of slicks on the ocean surface has also its specifics. Figures 6.24 and 6.25 show the panoramic image of the surface spot having a slick caused by a monomolecular film, and the panorama line marked by a white line in Figure 6.24. These data have been obtained with the help of SIR-C/X-SAR radar mounted on a space shuttle *Endeavour* working at 3 cm and decimetre waves with the nominal resolution 25m (Gade et al. 1998). A prominent feature in the image is the transition areas between the slick-covered and clean surface exceeding considerably (at least several times) the resolution cell nominal azimuthal size.

Figure 6.26 shows the numerically modelled panorama of the 2×2 km^2 SAR image of the windsea at the wind speed 8 m s^{-1} propagating at the angle of 20° against SAR platform flight direction (as earlier we assume $R/V = 120$s, $\theta_0 = 30°$, $\Delta_{0,\text{SAR}} = 7.5$m). The slick with the round shape of 500m diameter with well-defined borders and hydrodynamic contrast $K_{\text{slick}} = -10$ dB is situated in the centre of the panorama (by "hydrodynamic contrast" we understand the levelling extent of the Bragg waves inside the slick). Figure 6.27 displays the panorama line following the SAR flight direction along the slick diameter line. The imagery characteristic features are first the irregular (vague) borders of the slick (Figure 6.26) and, second, as before, a lot lower modulation of the signal intensity at the transition stage between slick-covered and clean surface (Figure 6.27) than it would have been should the nominal resolution $\Delta_{0,\text{SAR}} = 7.5$m specified in the formula play part. In compliance with the theory herein, the characteristic scale of this modulation along SAR flight direction is $2(R/V)\tilde{\sigma}_{\text{r}}$, which equals in our case to approximately 100 m.

Figure 6.24 SAR image $(2.5 \times 2.5\,\text{km}^2)$ of the sea surface in the presence of the slick caused by monomolecular organic film. The image was obtained from SIR-C/X, VV polarization. (Gade et al. 1998).

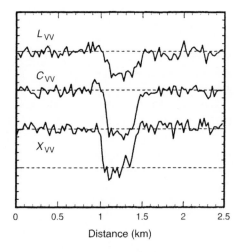

Figure 6.25 The row of the image given in Figure 6.24 (along the scan line shown as the white strip) for various frequency bands: $f_0 = 1.25\,\text{GHz}$ (L-band), $f_0 = 5.30\,\text{GHz}$ (C-band) and $f_0 = 9.60\,\text{GHz}$(X-band).

Hence we conclude that the estimate of the slick hydrodynamic contrast determined by its SAR image is lower if the slick dimensions along SAR platform flight direction is below $2(R/V)\hat{\sigma}_r$. This circumstance can have a significant negative impact in the case of space-borne detection of long and narrow-shaped oil spills caused by ship accidence, if

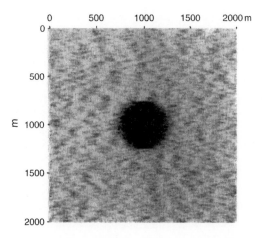

Figure 6.26 Modelled SAR image of the ocean surface in the presence of the slick of $-10\,\text{dB}$ hydrodynamic contrast. Fully developed windsea at wind speed $8\,\text{m s}^{-1}$; general direction of wave travel $\Phi_0 = 20°$.

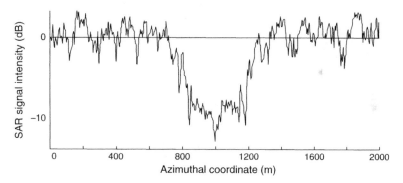

Figure 6.27 The horizontal row of the image given in Figure 6.28, passing through the slick centre.

radar carrier ground track crosses the slick transversely. At fairly intense waves the linear oil spill of the kind can be easily detected by eye, while most of automatic detections fail (Mercier and Girard-Ardhuin 2005).

6.8.3 SAR as a tool for measurements of the near-surface wind speed

As is known, the near-surface wind is measured from space with scatterometers and radar altimeters; today's algorithms (Karaev et al. 2002a) can restore the information with an accuracy up to about $1.5 - 1.7\,\text{m s}^{-1}$.

Physically, the function of scatterometers is based on the dependence of NRCS on the near-surface wind speed. The model function linking NRSC to local wind speed for probing at the angle φ_0 to the wind direction is given by

$$\sigma_0 = aU^\gamma(1 + b\cos\phi + c\cos 2\varphi_0) \qquad (6.166)$$

where a, b, c and γ are empirically determined coefficients that in general depend on radar wavelength, polarization and incidence angle.

The complex relation between NRCS and the wind speed vector is illustrated in Figure 6.28 (reproduced from Monaldo and Beal (2004)), where the three-dimensional plot of a function corresponding to microwave wind speed model CMOD-4 (Stofflen and Anderson 1997a, 1997b) for a 25° incident angle is represented. Clearly, one and the same NRCS value yields different data for both the wind speed U and the angle φ_0. For this reason, in scatterometers they use in parallel several radar beams directed differently against the platform flight direction. The main drawback of the space-borne scatterometer is its coarse resolution (30 – 50 km) totally inapplicable for off-coast regions which are characterized by very changeable spatial environmental characteristics.

Unlike a scatterometer, SAR has high (sub-kilometre) resolution and at the same time provides a panoramic survey of the ocean surface. Yet, SAR probes the surface at the single and a priori indeterminate angle to wind speed; hence the wind vector retrieval is

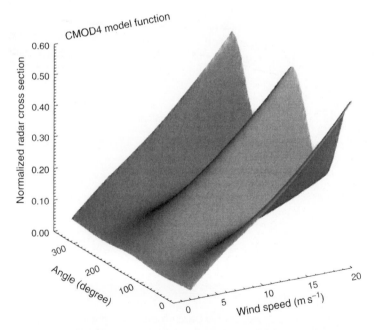

Figure 6.28 The CMOD4 geophysical model function relating wind speed and direction with respect to the radar normalized cross section. For this case, the incidence angle is 25°.

only by the NRCS value, even with the help of well-calibrated SAR, is generally unfeasible. However, in some particular cases it becomes possible to estimate wind speed direction on SAR image panorama with 180° ambiguity by some attributes, for instance, linear features aligned with the direction of the wind (wind row signatures) (Gerling 1986). Besides, one can use wind directions from model predictions. The well-illustrated examples of successful results of such approach can be found in Monaldo and Beal (2004).

As SAR builds a high-resolution image panorama, in this case unlike the scatterometer, we get the opportunity to identify and discriminate the changes of NRCS, which are caused by the strange (i.e. not associated with wind speed changes) reasons, sea ice, for instance.

The solution search to this problem can be greatly advanced by unification of space SAR and scatterometer (Furevik and Korsbakken 2000). The advantages of the combined use of SAR and scatterometer (the latter provides information on the wind direction) are also illustrated in Horstmann et al. (2003), where the bulk data from ERS-1, ERS-2 and ENVISAT, covering great ocean surface areas, displayed little disparity in wind speed values received with a coarse resolution scatterometer, on the one hand, and high-resolution SAR, on the other. One can expect that the synergistic fusion of data from both SAR and scatterometer will promote the essential advancement in the problem of wind field measurements in coastal areas as well.

Approaching the capacities of SAR in measurement of near-surface wind speed, it is reasonable to give another thought to SAR inherent capability of reducing the azimuthal component of the wave vector corresponding to the roughness image spectrum maximum. This phenomenon, though bears little effect on the flat swell due to the small value of the swell orbital velocity, becomes quite prominent in the case of wind waves (see Figure 6.1). A natural question comes up: Can't we determine the near-surface wind speed by the image spectrum peak shift, and if we can, what accuracy is achievable then? Surely, we have to somehow reduce the signal of the speckle noise before arriving at any definite conclusions.

Figure 6.29a–c gives the realizations of SAR signal spectra together with realizations of ocean wave spectra obtained with the help of numerical modelling on the basis of wind wave spectrum model (6.92) for three wind speed values: $U = 5, 10$ and $15 \, \mathrm{m \, s^{-1}}$. The waves were supposed to propagate azimuthally (or anti-azimuthally), each surface realization length was $2 \, \mathrm{km}$, SAR parameters set for the computations were $R/V = 120 \, \mathrm{s}, \theta_0 = 30°$ and $\Delta_{0,\mathrm{SAR}} = 7.5 \mathrm{m}$. We see that all the image spectra moved to the low spatial frequencies and the relative shift is

$$\Delta\dot{\kappa} = \frac{\kappa_0^{\mathrm{max}} - \kappa_{\mathrm{SAR}}^{\mathrm{max}}}{\kappa_0^{\mathrm{max}}} \tag{6.167}$$

where κ_0^{max} and $\kappa_{\mathrm{SAR}}^{\mathrm{max}}$ are the peak wave numbers in the roughness and image spectra, respectively, and decrease with the increase in wind speed. This is in good accordance with the developed theory. Indeed, in compliance with the computation results in Figure 6.6, at $\beta_\zeta > 0.2$, the condition $\Delta\dot{\kappa} > 0$ works; besides, with the lessening of β_ζ the value $\Delta\dot{\kappa}$ also decreases. At the same time, as Figure 6.9 displays, at the specified computation

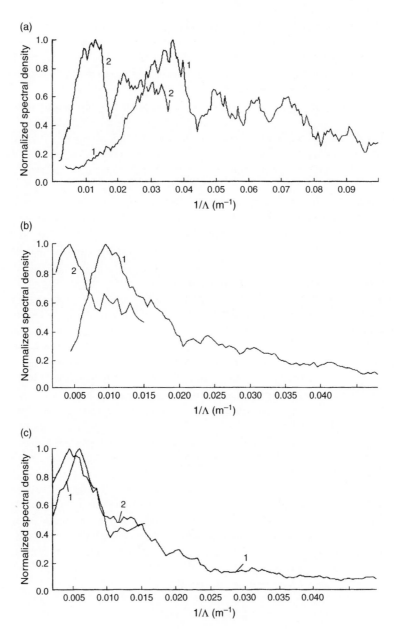

Figure 6.29 Surface elevation spectra (1) and the corresponding SAR image spectra (2) for various values of wind speed: $U = 5\,\mathrm{m\,s^{-1}}$ (a), $U = 10\,\mathrm{m\,s^{-1}}$ (b) and $U = U = 15\,\mathrm{m\,s^{-1}}$ (c).

parameters, first, $\beta_\zeta > 0.2$ at all given wind speed values, and second, in the range $5\,\mathrm{m\,s^{-1}} < U < 15\,\mathrm{m\,s^{-1}}$ with increase in the wind speed β_ζ decreases. Thus, in the specified wind speed range there occurs the spectral cut-off effect, and in accordance with Eqns (6.54) and (6.55) for azimuthally travelling waves

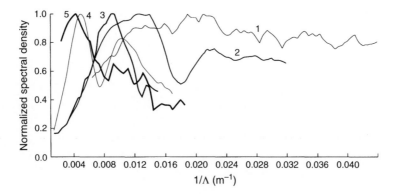

Figure 6.30 SAR image spectra of windsea for various values of wind speed.

$$\kappa_{\text{cut-off}} = \frac{\pi}{0.2} \left(\frac{R}{V} \right)^{-1} U^{-1} \sec\theta_0 \tag{6.168}$$

Figure 6.30 gives the image spectrum realizations of azimuthally travelling waves calculated at several values of wind speed. One can see that under low wind conditions ($U \leq 7\,\text{m}\,\text{s}^{-1}$) the peak positions considerably change every bar of $\Delta U = 1\,\text{m}\,\text{s}^{-1}$, but at $U \geq 10\,\text{m}\,\text{s}^{-1}$ the peak position of SAR image spectrum loses sensitivity to the wind speed. Yet, a more definite answer to the question above can be obtained only through an experiment.

– 7 –

Advanced radars and ocean surface imaging

Chapters 5 and 6 are devoted to conventional side-looking radars (RARs and SARs), operating at HH or VV polarizations and imaging the surface as the intensity (or amplitude) field of the reflected signal. Although the conventional SAR employs the signal phase, it does so only at the transitional stage of signal processing to achieve high resolution and eventually ocean imaging thus obtained outlines only the reflectivity of various surface elements, if we ignore the distorting effects caused by orbital velocities.

However the information on roughness collected by common imaging procedures based on the backscattered signal characteristics is by far inexhaustive. First, the RAR operating at co-polarized HH or VV signals does not practically see azimuthally travelling waves. Secondly, RAR is *basically* incapable of directly detecting surface velocities and sees them only in case of reflectivity variance brought about by currents, for example. As to SAR, both azimuthally travelling waves and surface currents become apparent in the SAR image of the ocean only indirectly, by means of velocity bunching (as a matter of fact, distorting) effects considered in Chapter 6. Besides, which is quite essential, roughness spectrum recovery only by energetic characteristics of the returned signals is hampered due to the poor knowledge of the MTF.

Overcoming the disadvantages is rendered possible by new imaging techniques of advanced, namely, interferometric and polarimetric radars. Respective technologies opening new perspective for radio-oceanography have been under active development for the past two decades.

7.1 SAR INTERFEROMETRY AND REMOTE SENSING OF THE OCEAN SURFACE

Earlier (see Sections 2.4 and 3.2.1) we talked about the significance of surface/sub-surface currents mapping carried out in coastal regions partly with HF radars. As for the open ocean, HF radar can reach rather far, though not always and everywhere, only under favourable ionosphere conditions. If the objective is the global and detailed mapping of ocean currents, the first and foremost tools then are airborne and space-borne microwave radars characterized by high resolution and insensitivity to the atmospheric conditions. The latter means that electromagnetic microwaves travel freely through atmosphere, clouds in particular; however, the atmospheric phenomena affecting the water surface in any way are displayed in radar imagery.

Ocean currents are often accompanied by roughness modulations, which can basically be detected with microwave imaging radar systems. Still the causes for roughness variances, besides currents, may vary from the near-surface wind speed fluctuations to stability effects dependent on the differences between the air and water upper layer temperatures, surfactant films and so on (see Section 2.4).

Obviously, it is hard to single out ocean currents among other numerous oceanic processes with the only help of radar image, having no extra environmental information. Therefore the solution has to be sought with airborne and space-borne microwave interferometric SAR (InSAR) which is designed mainly to assess the speed of ocean surface currents.

Since its inception (Goldstein and Zebker 1987, Goldstein et al. 1989), InSAR has been a promising and increasingly popular technique for remote sensing of the ocean surface and undersurface phenomena manifested on the surface due to the generated intrinsic velocity field created by them. Unlike conventional SAR, InSAR in its traditional configuration is equipped by two antennas mounted at baseline distance B from each other along the carrier track direction, and surveying the same surface spot (see Figure 7.1); such InSAR is termed along-track InSAR (AT-InSAR). AT-InSAR mechanism works by measuring the difference between the Doppler frequency phases of the signal reflected from the probed surface spot and received by the fore and then aft antenna consecutively over the time lag $\tau = B/V$ or $\tau = B/2V$, where V is the AT-InSAR platform velocity. (The time lag is B/V if both antennas are used to transmit and receive separate signals and the time lag is $B/2V$ if both antennas are used as receiving antennas for a signal originating from one of the antenna (Carande 1992).) After we measure the difference between phases ϕ over the time lag τ, the Doppler frequency of the backscattered signal is found:

$$f = \frac{\phi}{2\pi\tau} \tag{7.1}$$

and then the line-of-sight (i.e. radial) velocity of the scatterers on the surface:

$$v_{\text{surf}}^{\text{rad}} = \frac{\lambda f}{2\sin\theta_0} = \frac{\lambda\phi}{4\pi\tau\sin\theta_0} \tag{7.2}$$

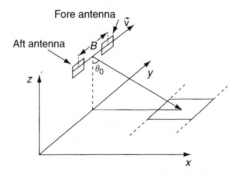

Figure 7.1 Schematic geometry of the AT-InSAR.

Here, λ is the wavelength of the incident electromagnetic radiation and θ_0 is the incidence angle.

As a matter a fact, the current velocity can also be measured with coherent microwave radar at an average frequency of the reflected signal Doppler spectrum. Yet, here, in the first place, if the sensing is performed in the open ocean, there arises the problem of finding the average frequency corresponding to the zero speed of the current. In the second place, Doppler spectrum is an average statistical characteristics reflecting the distribution of the line-of-sight velocity of the scatterers, weighted by their contributions to the backscattered power. Unlike this, using InSAR one can directly detect the Doppler frequency associated with each pixel of SAR image.

Using Eqns (7.1) and (7.2) one can write the conversation factor that relates InSAR phases to surface velocities and determine the accuracy of velocity measurements (Romeiser and Thompson 2000):

$$\frac{\Delta v_{\text{surf}}^{\text{rad}}}{\Delta \phi} = \frac{\lambda}{4\pi\tau\sin\theta_0} \; (\text{cm s}^{-1}\,\text{rad}^{-1}) = \frac{\lambda}{720\tau\sin\theta_0} \; (\text{cm s}^{-1}\,\text{deg}^{-1}) \qquad (7.3)$$

As we see from Eqn. (7.3), the measurement accuracy level of the current increases as the time lag τ used in computation of the phase difference ϕ becomes longer. However, τ cannot be infinitely large, not only because the distance B between AT-InSAR two antennas cannot be infinite, but also the difference between the phases is impossible to measure in principle over the temporal span τ exceeding the typical time spell of backscattered signal fluctuations, i.e. the time lag must not exceed the backscattered signal correlation time τ_{cor}. Thus, to appraise the accuracy with which the currents velocities are measured by AT-InSAR, τ_{cor} appears indispensable.

We consider the correlation function

$$B_{\text{INSAR}}(\tau) = \langle a_{\text{SAR(fore)}}(\vec{r},t)a^*_{\text{SAR(aft)}}(\vec{r},t+\tau)\rangle \qquad (7.4)$$

where $a_{\text{SAR(fore)}}$ and $a_{\text{SAR(aft)}}$ are the complex amplitudes of SAR signals from the fore and aft antennas; the asterisk denotes complex conjugation. Referring Eqn. (6.9) for SAR complex signal at the point $\vec{r}(x, y = Vt)$, we get

$$
\begin{aligned}
a_{\text{SAR(fore)}}(\vec{r},t) \propto \int dt_1 \exp\left[-2\left(\frac{t-t_1}{\Delta t}\right)^2\right] \\
\times \int_{\Delta \vec{r}} d\vec{r}_1\, m(\vec{r}_1,t_1)\xi(\vec{r}_1,t_1) \\
\times \exp\left\{2ik\left[(x_1-x)\sin\theta_0 - \zeta(\vec{r}_1,t_1)\right.\right. \\
\left.\left. + \frac{1}{2R}\left(y_1^2 - 2Vt_1(y_1-Vt) - V^2t^2\right)\right]\right\}
\end{aligned}
\qquad (7.5)
$$

Here, as before, Δt is the SAR integration time, $\Delta \vec{r}$ is the physical resolution cell part employed for the aperture synthesis (see Chapter 1), $\xi(\vec{r},t)$ denotes "standard" ripples homogeneously spread on the large-scale surface $\zeta(\vec{r}, t)$ formed by large waves, and the multiplier $m(\vec{r}, t)$ describes the modulation of the reflected field amplitude, both geometrical and hydrodynamic.

The signal generated by the aft antenna in the same point over the time lag τ is as follows:

$$
a_{\text{SAR(aft)}}(\vec{r}, t + \tau) \propto \int dt_2 \exp\left[-2\left(\frac{t - t_2}{\Delta t}\right)^2\right] \int_{\Delta \vec{r}} d\vec{r}_2 \, m\left(\vec{r}_2, t_2 + \tau\right)\xi\left(\vec{r}_2, t_2 + \tau\right)
$$

$$
\times \exp\left\{2ik\left[(x_2 - x)\sin\theta_0 - \zeta\left(\vec{r}_2, t_2 + \tau\right)\right.\right.
$$

$$
\left.\left. + \frac{1}{2R}\left(y_2^2 - 2Vt_2(y_2 - Vt) - V^2t^2\right)\right]\right\} \tag{7.6}
$$

Two complex signals correlation are as follows:

$$
B_{\text{InSAR}}(\tau) = \left\langle a_{\text{SAR(fore)}}(\vec{r}, t) a_{\text{SAR(aft)}}^*(\vec{r}, t + \tau)\right\rangle
$$

$$
\propto \left\langle \iint dt_1 \, dt_2 \, \exp\left[-2\left(\frac{t - t_1}{\Delta t}\right)^2\right]\exp\left[-2\left(\frac{t - t_2}{\Delta t}\right)^2\right]\right.
$$

$$
\times \iint_{\Delta \vec{r}} d\vec{r}_1 \, d\vec{r}_2 \, m\left(\vec{r}_1, t_1\right)m^*\left(\vec{r}_2, t_2 + \tau\right)B_\xi\left(\vec{r}_1 - \vec{r}_2, t_1 - t_2 - \tau\right)
$$

$$
\times \exp\left\{2ik\left[(x_1 - x_2)\sin\theta_0 - \left[\zeta\left(\vec{r}_1, t_1\right) - \zeta\left(\vec{r}_2, t_2 + \tau\right)\right]\cos\theta_0\right]\right\}
$$

$$
\left. \times \exp\left[\frac{ik}{R}\left(y_1^2 - y_2^2 - 2V(t_1y_1 - t_2y_2) + 2V^2t(t_1 - t_2)\right)\right]\right\rangle \tag{7.7}
$$

Angular brackets indicate averaging process over large-scale roughness realizations; averaging over small "standard" ripples is assumed to have already been performed (this is supported by the correlation function B_ξ in the integrand of Eqn. (7.7)).

As mentioned before (see Section (5.2)) we introduce variables

$$
\vec{r}_1 - \vec{r}_2 = \rho', \qquad \vec{r}_1 - \vec{r}_2 = 2\vec{r}\,'
$$

$$
t_1 - t_2 = \tau', \qquad t_1 - t_2 = 2\tau' \tag{7.8}
$$

As the correlation function of the ripples B_ξ is fast to decrease on the spatio-temporal scale, which is small against the characteristic period and wavelength of large roughness, then within the interval where the integrand of Eqn. (7.7) is sufficiently non-zero, we can assume

$$
m(\vec{r}_1, t_1) = m(\vec{r}_2, t_2) = m(\vec{r}\,', t') \tag{7.9}
$$

$$
\zeta(\vec{r}_1, t_1) - \zeta(\vec{r}_2, t_2) = \frac{\partial\zeta}{\partial x}(\vec{r}\,', t')\rho'_x + \frac{\partial\zeta}{\partial y}(\vec{r}\,', t')\rho'_y + \frac{\partial\zeta}{\partial t}(\vec{r}\,', t')\tau' \tag{7.10}
$$

where ρ'_x and ρ'_y are the components of the vector $\vec{\rho}\,'$. Besides, we assume the ground range size Δx of a physical element to be small against the characteristic surface wavelength. Then Eqn. (7.7) can be written as

$$
\begin{aligned}
B_{\text{InSAR}}(\tau) \propto \Bigg\langle & \int\limits_{\Delta y} \mathrm{d}y \int \mathrm{d}t' \left| m(\vec{r}\,',t') \right|^2 \exp\left[-4\left(\frac{t-t'}{\Delta t} \right)^2 \right] \exp\left[-2\mathrm{i}k\tau \frac{\partial \zeta}{\partial t}(\vec{r}\,',t') \cos\theta_0 \right] \\
& \times \int\limits_{-\infty}^{\infty} \mathrm{d}\tau' \exp\left\{ -\frac{\tau'^2}{(\Delta t)^2} - 2\mathrm{i}k\left[\frac{\partial \zeta}{\partial t}(\vec{r}\,',t') \cos\theta_0 - \frac{V}{R}(Vt - y') \right]\tau' \right\} \\
& \times \int\limits_{-\infty}^{\infty} \mathrm{d}\vec{\rho}\,' B_\xi\left(\vec{\rho}\,', \tau'-\tau \right) \exp\left\{ 2\mathrm{i}k \begin{bmatrix} \left(\sin\theta_0 - \frac{\partial \zeta}{\partial x}(\vec{r}\,',t') \cos\theta_0 \right)\rho'_x \\ + \left(\frac{y'-Vt'}{R} - \frac{\partial \zeta}{\partial y}(\vec{r}\,',t') \cos\theta_0 \right)\rho'_y \end{bmatrix} \right\} \Bigg\rangle
\end{aligned}
\tag{7.11}
$$

Ignoring the small variations in resonance wave number within the physical resolution cell, i.e. assuming

$$
\vec{\kappa}_{\text{res}}(\vec{r}\,',t') = \{-2k \sin\theta_0, 0\}
\tag{7.12}
$$

we write the last integral in Eqn. (7.11) as

$$
\begin{aligned}
& \int\limits_{-\infty}^{\infty} \mathrm{d}\vec{\rho}\,' \, B_\xi\left(\vec{\rho}\,',\tau'-\tau \right) \exp\left(2\mathrm{i}k \sin\theta_0 \cdot \rho'_x + 0 \cdot \rho'_y \right) \\
& = 4\pi^2 \hat{W}_\xi\left(\vec{\kappa}_{\text{res}} \right) \exp\left\{ -\mathrm{i}\left[\Omega_{\text{res}} + \vec{\kappa}_{\text{res}} \vec{v}_{\text{surf}}(\vec{r}\,',t') \right](\tau'-\tau) \right\}
\end{aligned}
\tag{7.13}
$$

where $\hat{W}_\xi(\vec{\kappa})$ is the spatial spectrum of "standard" ripples and $\Omega_{\text{res}} = (\kappa_{\text{res}}g)^{1/2}$ is the inherent frequency of the ripple resonance component; the water particles move on the surface $\vec{v}_{\text{surf}}$ with the velocity that comprises orbital velocity, wind drift, Stokes velocity component and current velocity. When computing Eqn. (7.13) we took into account that

$$
B_\xi\left(\vec{\rho}\,',\tau'-\tau \right) = \iint \mathrm{d}\vec{\kappa} \, \mathrm{d}\omega \, \Psi_\xi\left(\vec{\kappa},\omega \right) \exp\left\{ \mathrm{i}\left[\vec{\kappa} \vec{\rho}\,' - \omega(\tau'-\tau) \right] \right\}
\tag{7.14}
$$

$$
\Psi_\xi\left(\vec{\kappa}_{\text{res}},\omega \right) = \hat{W}_\xi^+\left(\vec{\kappa}_{\text{res}} \right) \delta\left(\omega - \left(\Omega_{\text{res}} + \vec{\kappa}_{\text{res}} \vec{v}_{\text{surf}} \right) \right)
\tag{7.15}
$$

We wrote the second equation assuming (for simplicity) that the ripple spectrum lacks the components negatively projected onto the wind direction vector; besides, to make the calculations more explicit, we assume $\vec{k}\,\vec{U} < 0$.

Integration over t' becomes easy after we assume that $\Delta t \ll T_0$, where T_0 is the roughness characteristic temporal scale; just as easy is the integration over τ'. Performing the calculations analogous to those in Section 6.3, we find

$$
B_{\text{InSAR}}(\tau) \propto \left\langle \int dy' \, \sigma_0(x, y') \exp\left[-ikv_{\text{surf}}^{\text{rad}}(x, y')\tau\right] \right.
$$
$$
\left. \times \exp\left\{ -\frac{\pi^2}{\Delta_{0,\text{SAR}}^2} \left[Vt - y' - \frac{R}{V} v_{\text{surf}}^{\text{rad}}(x, y') \right] \right\} \right\rangle
\tag{7.16}
$$

We divide the surface radial velocity into constant and variable components:

$$
v_{\text{surf}}^{\text{rad}} = v_0 + \tilde{v}
\tag{7.17}
$$

The constant component includes all the velocities listed above, apart from the orbital value. We average σ_0 and the rest of the integrand in Eqn. (7.16) separately (we always did so before without any serious impact on the computation results), considering the variable (i.e. orbital) velocity obeys Gaussian distribution with the variance σ_{rad}^2:

$$
B_{\text{InSAR}}(\tau) \propto (\sigma_0(x, y')) \exp\left(-2ikv_0\tau - 2k^2\sigma_{\text{rad}}^2 \frac{\Delta_{0,\text{SAR}}^2}{\Delta_{\text{SAR}}^2} \tau^2 \right)
$$
$$
\times \int dy' \exp\left\{ -\frac{\pi^2}{\Delta_{\text{SAR}}^2} \left(y - y' - \frac{R}{V} v_0 \right)^2 - i\frac{4\pi^2\sigma_{\text{rad}}^2}{\Delta_{\text{SAR}}^2} \frac{R}{V} \left(y - y' - \frac{R}{V} v_0 \right) k\tau \right\}
\tag{7.18}
$$

where $\Delta_{\text{SAR}}^2 = \Delta_{0,\text{SAR}}^2 + 2\pi^2(R/V)^2\sigma_{\text{rad}}^2$. Integrating the expression we get

$$
B_{\text{InSAR}}(\tau) \propto \exp\left(-2i\vec{k}\,\vec{v}_0\tau \right) \exp\left(-2k^2\sigma_{\text{rad}}^2\tau^2 \right)
\tag{7.19}
$$

Evidently, $B_{\text{InSAR}}(\tau)$ basically is similar to the correlation function (4.34) of electromagnetic field backscattered from the sea surface. The differences are not crucial: Eqn. (7.19) takes into account only one of the two ripple resonance components (as discussed above), besides, surface speed constant component here includes, apart from its proper ripple phase velocity, wind drift, Stokes velocity and the speed of the current.

One might wonder why we have obtained the expression for moving radar correlation function that precisely, to the constant coefficient, matches the one received in Chapter 3 for the stationary tool. Indeed, we stated there (see Eqn. (4.45)) that at a fairly high speed of radar platform the width of Doppler spectrum and, consequently, of the correlation function do not depend on the roughness any longer. The answer is quite simple. As a matter of fact, unlike Chapter 3, we analyse here SAR, whose signal due to the signal matched filtering (1.4) lacks (or, to be more exact, offsets) the fluctuations induced by the carrier motion, which in RAR prevail over roughness-generated fluctuations; this point has been considered in Chapter 3.

Therefore,

$$\tau_{\text{cor}} = \frac{\lambda}{2\sqrt{2}\pi\sigma_{\text{rad}}} \qquad (7.20)$$

Considering that

$$\sigma_{\text{rad}} \approx \left(\cos^2\theta_0 + \cos^2\varphi_0 \sin^2\theta_0\right)^{1/2}\sigma_{\text{orb}} \qquad (7.21)$$

under the windsea conditions $\sigma_{\text{orb}} = \sigma_{\text{vert}} \approx 6.8 \times 10^{-2} U$ (the last approximate equation has been obtained in line with Eqn. (2.23)), we finally get

$$\tau_{\text{cor}} \approx \frac{1.65\lambda}{\left(\cos^2\theta_0 + \cos^2\varphi_0 \sin^2\theta_0\right)^{1/2}U} \qquad (7.22)$$

Remember that U is the near-surface wind speed and φ_0 is the angle between wind direction and radar look direction.

In the case of wavelength $\lambda = 3.2$ cm, $U = 10$ m s^{-1} and $\varphi_0 = 45°$,

$$\tau_{\text{cor}} \approx \begin{cases} 6.7 \text{ ms}, & \theta_0 = 30° \\ 5.7 \text{ ms}, & \theta_0 = 60° \end{cases} \qquad (7.23)$$

Notably, Eqn. (7.22) gives the results τ_{cor} quite close to those in Romeiser and Thompson (2000), obtained from Doppler spectrum numerical modelling of the microwave field backscattered from the sea surface. It is perfectly natural, as Fourier transformation of Eqn. (7.19) results exactly in the Doppler spectrum. SAR inherent features are manifested in the generation at every instant of a signal backscattered by the area with the average azimuthal size $\Delta_{\text{SAR}} \pm (R/V)\hat{\sigma}_{\text{orb}}^{\text{rad}}$, shifted azimuthally against $y = Vt$ by the distance $(R/V)v_0$ along or back the SAR platform path depending on the sign v_0. These features are studied in detail in Chapter 6.

Now we can appraise the measurement accuracy of the surface radial velocity. As Romeiser and Thompson (2000) remark, the measuring accuracy of the phase difference $\Delta\phi$ in modern InSAR is approximately 1°. Substituting the values of $\tau = 3$ ms and $\Delta\phi = 1°$ into Eqn. (7.3), we find for $\lambda = 3.2$ ms,

$$\Delta v_{\text{surf}}^{\text{rad}} = \begin{cases} 3.0 \text{ cm s}^{-1}, & \theta_0 = 30° \\ 1.7 \text{ cm s}^{-1}, & \theta_0 = 60° \end{cases} \qquad (7.24)$$

Hence we can conclude that phenomena such as wind drift and Stokes drift current (see Section 2.3), in principle, remain within the limits of speed measuring accuracy range provided by AT-InSAR, and should be considered when getting the speed of the currents.

Evidently, at $\phi \geq 2\pi$ it becomes difficult to measure the speed of the current, when it is impossible to discern v and $v + \Delta v_{\mathrm{amb}}$ by InSAR imagery. Expression (7.3) gives

$$\Delta v_{\mathrm{amb}} = \frac{\lambda}{2\tau \sin \theta_0} \qquad (7.25)$$

which yields $\Delta v_{\mathrm{amb}} \approx 6 \text{ m s}^{-1}$ for $\lambda = 3.2 \text{ cm}, \tau = 3 \text{ ms}$ and $\theta_0 = 30°$, i.e. the ambiguity problem is not a crucial one.

When processing the sensing data obtained with AT-InSAR, it is necessary, first, to filter variable orbital velocities. The filtration is done by averaging AT-InSAR image over quite a number of pixels. Besides, we have to subtract the phase velocity of the ripple resonance component as well as wind drift and Stokes drift current velocities from the remaining constant component of the surface speed, which requires measuring wind speed and direction. In contrast to the simplified case analysed above, generally there are two Bragg waves (approaching and receding) with an unknown beforehand intensity ratio on the surface, and it takes at least two-frequency AT-InSAR (Kim et al. 2003) to filter out the phase velocity of the Bragg waves.

The first in situ data on space AT-InSAR can be found in Romeiser et al. (2005). The typical design AT-InSAR described above does not measure the entire speed of the current, but rather the line-of-sight (radial) component of velocity vector. To measure the velocity vector, AT-InSAR carrier has to move along two perpendicular directions above the surveyed surface spot. This challenge of conventional InSAR has been overcome by more advanced SAR, namely dual-beam InSAR (DB-InSAR). The idea of DB-InSAR was originally suggested by Rodriguez et al. (1995) and developed by Frasier and Camps (2001). Then the dual-beam interferometer was eventually designed and built by the University of Massachusetts (USA) (Farquharson et al. 2004).

DB-InSAR combines two AT-InSARs, one squinted θ_{s1} forward from broadside and the other θ_{s2} aft (see Figure 7.2 quoted from Toporkov et al. 2005). Thus, every scene in

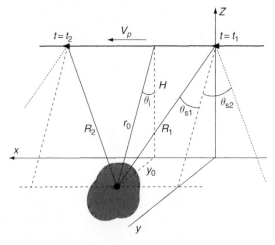

Figure 7.2 Schematic geometry of the DB-InSAR.

the swath is observed from two different directions, which allows reconstruction of the velocity vector. In accordance with Toporkov et al. (2005), the surface velocity vector along-track and cross-track components can be found as

$$v_{\text{surf}}^{\text{along-track}} = \frac{u_1 \cos \theta_{s2} - u_2 \cos \theta_{s1}}{\sin(\theta_{s1} - \theta_{s2})} \tag{7.26a}$$

$$v_{\text{surf}}^{\text{cross-track}} = \frac{u_2 \sin \theta_{s1} - u_1 \sin \theta_{s2}}{\sin(\theta_{s1} - \theta_{s2}) \sin \theta_0} \tag{7.26b}$$

where θ_0 is the incidence angle, and u_1 and u_2 are the radial velocities observed by squinted fore and aft InSARs, respectively (the aft-looking squint angle is assumed to be negative).

In Frasier and Camps (2001), the analysis of the influence of DB-InSAR platform attitude and velocity errors has been carried out, and in Toporkov et al. (2005) the results of first measurements of surface current velocity vector are given. These results show that DB-InSAR produces reasonable and consistent single-pass estimates of full surface velocity vector. The observed flow patterns correlate well with bathymetric features inferred from aerial photography, and measurements obtained from different flight tracks generally show consistency.

Another modification of the InSAR was analysed in Schulz-Stellenfleth and Lehner (2001). As opposed to discussed above along-track InSAR, the attention here is focused on single-pass across-track InSAR, which uses two antennas, transmitting and receiving radar signals in turn (Figure 7.3). Each antenna acquires a complex SAR image denoted as

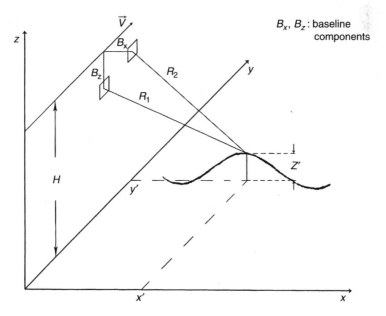

Figure 7.3 Schematic geometry of the across-track InSAR.

I_1 and I_2, respectively. The across-track SAR complex image is based on the interferogram defined as

$$I_i(\vec{r}) = \left\langle I_1(\vec{r}) I_2^*(\vec{r}) \right\rangle \tag{7.27}$$

where the asterisk denotes complex conjugation and angle brackets indicate averaging over ripples and sub-resolution velocities.

Across-track InSAR deploys such an antenna set pattern, because the difference between the phases received by the two antenna signals backscattered from the arbitrary point $\vec{r}' = \{x', y'\}$ depends on the elevation $z' = \zeta(\vec{r}', t)$ of the large-scale ocean surface. More specifically, the difference is $2k\Delta R$, where k is the electromagnetic radar wave number and ΔR is the difference between the distances from the point $\vec{r}$ to each antenna:

$$\Delta R = R_1 - R_2 \tag{7.28a}$$

$$R_1(\vec{r}', t) = \left[(H - B_z - z'(t))^2 + (y' - y)^2 + x'^2 \right]^{1/2} \tag{7.28b}$$

$$R_2(\vec{r}', t) = \left[(H - z'(t))^2 + (y - y')^2 + (x' - B_x)^2 \right]^{1/2} \tag{7.28c}$$

Making the same computations as for the along-track InSAR above, and averaging over the realizations of small ripples and sub-resolution velocities, we find for the across-track SAR interferogram,

$$I_i(x, y) \propto \int dy'\, \sigma_0(x, y')\, \exp[2ik\, \Delta R(x, y')]\, \exp\left\{ -\frac{\pi^2}{\Delta_{SAR}} \left[y - y' - \frac{R}{V}\, \tilde{v}_{rad}(x, y') \right]^2 \right\} \tag{7.29}$$

where Δ_{SAR} is the SAR resolution cell smeared because of sub-resolution orbital velocities, and $\tilde{v}_{rad}$ the orbital velocity radial component caused by large-scale ocean waves. Thus, the complex across-track InSAR image phase contains immediate information on the large-scale roughness elevations; therefore, this kind of image is less dependent on the scarcely explored MTF than the conventional SAR image depicting only surface reflectivity. At the same time, the way formula (7.28) is structured proves that the across-track InSAR image contains the same velocity bunching effects, namely azimuthal cut-off as well as the displacement of the spectral peaks, as in the conventional SAR image. The fact is also supported by the experiment results obtained with airborne across-track InSAR (Schulz-Stellenfleth et al. 2001). The experimental data also indicate, which is quite important, that the signal-to-noise ratio (SNR) of bunched InSAR spectra is shown to be about 5–10 dB higher than the SNR of conventional SAR intensity image spectra.

7.2 POLARIMETRIC RADARS AND THE PROBLEM OF REMOTE SENSING OF THE OCEAN SURFACE

Another information source on the roughness structure is polarization variance in radar ocean backscattering.

As we already know, RAR working on co-polarized HH or VV signals does not see the surface slopes in the plane perpendicular to the plane of incidence. SAR is capable of seeing these slopes, yet not directly but due to orbital velocities caused by azimuthally travelling waves. As we shall further learn, exploiting other combinations of transmitted and received signal polarization allows both RAR and SAR to see above-mentioned slopes directly.

We introduce the notion of polarization orientation angle φ_{pol}, which is illustrated in Figure 7.4 (from Lee et al. 2000); for an elliptically polarized signal, ellipticity angle χ is also displayed here.

Let an electromagnetic wave, generally elliptically polarized, fall onto the quasi-flat rough plane slanting perpendicularly to the incidence plane ($\alpha_x = 0$, $\alpha_y \neq 0$). In this case, the incident field polarization towards the tilted facet differs compared to the polarization towards a non-tilted one. It results in the change of backscatter conditions and, in its turn, the change of the polarization orientation angle, as well as the ellipticity angle. The angle φ_{pol} can be obtained by rotating a horizontal- and vertical-based antenna configuration by an angle such that return is a maximum.

In general case ($\alpha_x \neq 0$, $\alpha_y \neq 0$), the induced polarization orientation angle shift $\Delta\varphi_{pol}$ is given by the following formula (Lee et al. 2000):

$$\tan\Delta\varphi_{pol} = \frac{\tan\alpha_y}{-\tan\alpha_x \cos\theta_0 + \sin\theta_0} \quad (7.30)$$

This equation shows that the polarization orientation angle shift is mainly induced by azimuth slope, and in general case it is a function of the range slope and the incidence angle. Taking into account smallness of α_x and α_y one can write down

$$\Delta\varphi_{pol} \approx \frac{\alpha_y}{\sin\theta_0} \quad (7.31)$$

and in this case the dependence on range slope is practically absent.

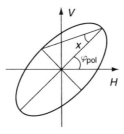

Figure 7.4 Polarization ellipse, φ_{pol} is the orientation angle between $[0, \pi)$, and χ is the ellipticity between $[-\pi/4, \pi/4]$.

Thus, looking for the polarization orientation shift one can see the azimuthally travelling waves, whereas RAR working on co-polarized VV or HH signal cannot (see Section 3.2.2). The capacity of RAR to detect azimuthally travelling waves, receiving a cross-polarized signal, is proved by also by formulae (5.1)–(5.3) of Valenzuela (1978) for the local cross section of backscatter from sea surface element; here they are in full:

$$\sigma_{0,\text{HH}} = 4\pi k^4 \cos^4 \theta_0 \left| \left(\frac{\sin(\theta_0 - \alpha_x) \cos \alpha_y}{\sin \theta_0} \right)^2 g_{\text{HH}} + \left(\frac{\sin \alpha_y}{\sin \theta_0} \right)^2 g_{\text{VV}} \right|^2$$
$$\times \hat{W}_\xi \left(-2k \sin(\theta_0 - \alpha_x), \ -2k \cos(\theta_0 - \alpha_x) \sin \alpha_y \right) \tag{7.32a}$$

$$\sigma_{0,\text{VV}} = 4\pi k^4 \cos^4 \theta_0 \left| \left(\frac{\sin(\theta_0 - \alpha_x) \cos \alpha_y}{\sin \theta_0} \right)^2 g_{\text{VV}} + \left(\frac{\sin \alpha_y}{\sin \theta_0} \right)^2 g_{\text{HH}} \right|^2$$
$$\times \hat{W}_\xi \left(-2k \sin(\theta_0 - \alpha_x), -2k \cos(\theta_0 - \alpha_x) \sin \alpha_y \right) \tag{7.32b}$$

$$\sigma_{0,\text{VH}} = \sigma_{0,\text{HV}} = 4\pi k^4 \cos^2 \theta_0 \left(\frac{\sin(\theta_0 - \alpha_x) \sin \alpha_y \cos \alpha_y}{\sin^2 \theta_0} \right)^2 |g_{\text{VV}} - g_{\text{HH}}|^2$$
$$\times \hat{W}_\xi \left(-2k \sin(\theta_0 - \alpha_x), \ -2k \cos(\theta_0 - \alpha_x) \sin \alpha_y \right) \tag{7.32c}$$

Here we have added the expression $\sigma_{0,\text{VH}} = \sigma_{0,\text{HV}}$ to Eqn. (3.36) and switched from the grazing angle ψ_0 to the incidence angle $\theta_0 = (\pi/2) - \psi_0$. In Eqn. (3.12) for g_{HH} and g_{VV} we have also to make the transition from the grazing angle to the incidence angle.

Notably, as we can see from the cited above expressions, operating at cross-polarized signals is more energy demanding in comparison with co-polarized one, because $\sigma_{0,\text{VH}} = \sigma_{0,\text{HV}} \ll \sigma_{0,\text{VV}}, \ \sigma_{0,\text{HH}}$.

Theoretical work (He et al. (2004)) analyses the case of radar transmitting a linearly polarized (*but not obligatory H or V*) signal and receiving a co-polarized one. The authors of cited work have obtained the RAR MTF in the form

$$T = T_{\text{tilt}} + T_{\text{pol}} = A_{\text{tilt}} i \kappa_x - A_{\text{pol}} i \kappa_y \tag{7.33}$$

where both A_{tilt} and A_{pol} depend on the polarization orientation angle φ_{pol} of transmitted signal as well as on the incidence angle and water dielectric constant ε. Without writing a rather cumbersome formulae for A_{tilt} and A_{pol}, we shall still mention that at $\varphi_{\text{pol}} = 0°$ or $\varphi_{\text{pol}} = 90°$ (i.e. at HH or VV polarization) $T_{\text{pol}} = 0$, and T_{tilt} reduces to the traditional tilt MTF discussed in Section 3.2.2.

Figure 7.5 (quoted from He et al. (2004)) shows dimensionless forms for tilt MTF A_{tilt} and polarization orientation MTF A_{pol} for different polarization orientation angles, plotted as a function of incidence angle. This figure demonstrated that larger incidence angles result in larger polarization orientation modulation, for all linear polarizations. Generally, the curves of Figure 7.5 illustrate the results providing the basis for improvement to the radar measurements of azimuthally travelling ocean waves, as suggested by Schuler and Lee (1995).

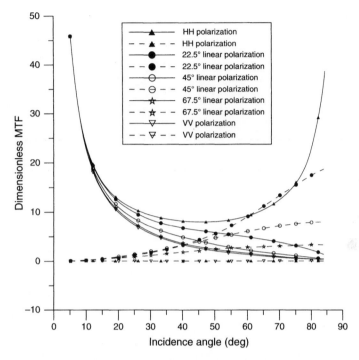

Figure 7.5 Dimensionless tilt MTF (solid lines) and polarization orientation MTF (dashed lines) as a function of incidence angle, for different polarizations (He et al. 2004).

To apply these computations to SAR, we have to insert into Eqn. (6.30) the NCRS dependent on the polarization orientation angle of transmitted signal, whose fluctuations are defined by RAR MTF (7.33). In Figure 7.6a–d we find the results of numerical simulation of the model input ocean wave spectrum, the corresponding conventional and linear-polarimetric RAR image spectra as well as the linear-polarimetric SAR image spectrum, all carried out in He et al. (2004). It is evident from Figure 7.6 that the polarization orientation modulation enhances our ability to measure azimuthal ocean waves. The comparison of simulated linear-polarimetric SAR image spectra obtained for azimuthally and anti-azimuthally travelling waves (Figure 7.7) reveals certain differences between them, which indicates the capability of resolving 180° directional ambiguity at one-look working regime.

The technique for measuring ocean surface slopes and wave spectra using polarimetric SAR data has also been developed in Schuler et al. (2004). Wave-induced perturbations of the orientation angle are used to sense the wave slopes. This technique forms a mean of using polarimetric SAR image data to make measurements of either ocean wave slopes or directional wave spectra.

It is to underline specifically that polarimetric RAR and SAR sense the surface slopes through polarization orientation angle immediately and not indirectly as through MTF and velocity bunching. We can expect then that the retrieval of the ocean spectra from polarimetric radar data will be more effective than from convention radars.

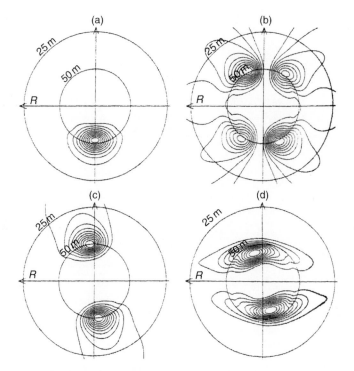

Figure 7.6 Input ocean spectrum (a), convention RAR spectrum (b), linear-polarimetric RAR spectrum (c) and linear-polarimetric SAR spectrum (d) (He et al. 2004).

Another vivid proof of polarimetric SAR fine performance in oceanography is in Schuler et al. (2003). This paper describes the investigation, which has been carried out on the use of polarization orientation angles to remotely sense ocean wave slope distribution changes caused by wave–current interactions. The current-induced asymmetry of wave slopes creates a mean slope that is manifested as a mean orientation angle.

As noted in Schuler et al. (2003), for many scattering cases, the maximum co-polarization response angular rotation can be measured quite sensitively. The ocean is a scattering case for which the maximum co-polarization response can be measured accurately. After sufficient averaging the orientation angle can be measured, using the co-polarization maximum, with an accuracy much better than $1°$.

The conventional intensity image of internal wave intersecting packets in the New York Bight (airborne SAR, L-band, VV data) is represented in Figure 7.8. The arrow indicates the propagation direction for the chosen study packet (within the dashed lines). The angle α relates the SAR/packet coordinates. Figure 7.9 is the orientation angle image of the same wave packets; the area within the wedge (dashed lines) was studied.

A profile of orientation angle perturbation caused by internal wave study packet is given in Figure 7.10a. The values are obtained along the propagation vector line of Figure 7.8. Figure 7.10b gives a comparison of the orientation angle profile (solid line) and a normalized backscatter intensity profile (dot–dash line) along the same interval.

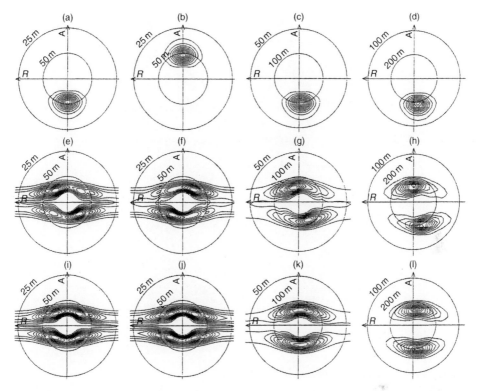

Figure 7.7 Input ocean wave spectra with the peak wavelength of (a) 50 m, (b) 50 m, (c) 100 m and (d) 200 m. Wave directions are anti-azimuthal in (a), (c) and (d), and azimuthal in (b). Corresponding polarimetric SAR spectra are in (e), (f), (g), and VV polarization conventional SAR spectra in (i), (j), (k) and (l) (He et al. 2004).

However, the angle changes are not produced by all types of ocean surface features. The orientation angle changes have been successfully used in the cited work to discriminate against other ocean features, such as surfactant slicks, that are present in the same scene. In Figure 7.11a the conventional image showing surfactant slicks (black scars) is given. Figure 7.11b is an orientation angle image of the same area. The slicks have been suppressed by the orientation angle processing technique.

The suppression mentioned above is caused by the fact that the polarization orientation angle depends only on large-scale surface slopes but not on the characteristics of small-scale ripples, which are sensitive to surfactant films. However if we are not talking about discrimination, but have in view slick detection (primary importance task we have repeatedly pointed out before), the polarization capacity of the radar signal should be deployed otherwise.

In Fuks and Zavorotny (2007) the polarization dependence of radar contrast for sea surface oil slicks has been considered. As we know, radar signal polarization dependence is defined by large-scale roughness, and in absence of large waves the slick radar contrast equals its hydrodynamic contrast (i.e. the intensity contrast of the ripple resonance component, to be exact) independent of the signal polarization. In real life, i.e. under

-1.0°   0°   +1.0°
Orientation angle

Figure 7.8 Airborne SAR (L-band, VV polarization) image of internal wave intersecting packets in the New York Bight. The arrow indicates the propagation direction for the chosen study packet (within the dashed lines). The angle α relates the SAR/packet coordinates (Schuler et al. 2003).

rough sea conditions, the dependence of slick radar contrast on the probing signal polarization is observed.

Using the two-scale scattering model, Fuks and Zavorotny (2007) have obtained the theoretical results that can be summarized as follows:

1. The contrast in the radar image of sea surface oil slicks obtained at HH and HV polarizations significantly exceeds that at VV polarization.
2. The contrast dependence on polarization increases as the grazing angle decreases.
3. The contrast increases as both wind speed and the spectral range occupied by the slick (slick cut-off spectral length) increase.
4. The polarization-dependent part of the contrast increases as the radar frequency increases: at K_u band it exceeds the one at C band.
5. The non-Gaussian PDF of slopes causes the difference between upwind and downwind contrasts.

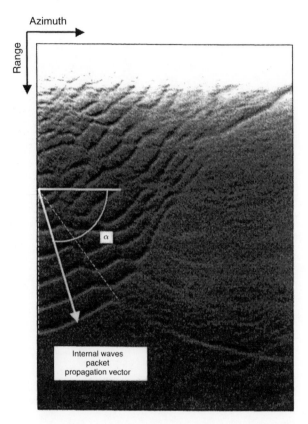

Figure 7.9 Orientation angle image of the internal wave packets in the New York Bight. The area within the wedge (dashed lines) was studied (Schuler et al. 2003).

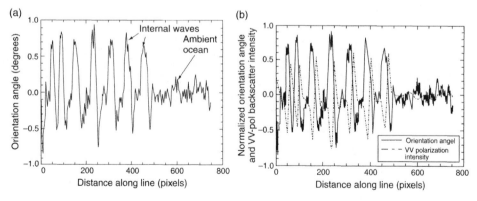

Figure 7.10 (a) Orientation angle value profile along the propagation vector for the internal wave study packet of Figure 7.8 and (b) a comparison of the orientation angle profile (solid line) and a normalized backscatter intensity profile (dot–dash line) (Schuler et al. 2003).

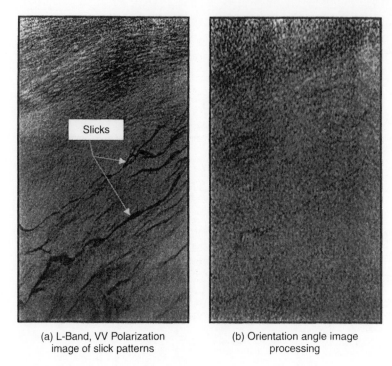

(a) L-Band, VV Polarization
image of slick patterns

(b) Orientation angle image
processing

Figure 7.11 The image of (a) is L-band VV polarization data showing surfactant slicks (black scars) in the New York Bight, and (b) is an orientation angle image of the same area. One can see that the slicks have been suppressed in (b) with the orientation angle processing technique (Schuler et al. 2003).

6. The contrast is azimuth-dependent: at VV it monotonically decreases, and at HH it increases when the azimuth direction varies from downwind to upwind. The contrast at cross-polarization (HV) increases from downwind to upwind, but has a minimum near the cross-wind direction.

The information presented in this section makes it clear that using polarimetric radars significantly boosts the potentialities of radar remote sensing of the ocean. Moreover, obviously, even more cardinal advancement can be reached if advantages provided by radar interferometry and polarimetry are joined in a high-resolution SAR (Boerner 2007).

Appendix A

Let the two values $a(\vec{r}, t)$ and $b(\vec{r}, t)$ be roughness linear functions, i.e.

$$a(\vec{r},t) = \int d\vec{\kappa}\, T_a(\vec{\kappa}) A_\zeta(\vec{\kappa}) \exp\left[i(\vec{\kappa}\vec{r} - \Omega(\kappa)t)\right] + \text{c.c.} \qquad (\text{A.1})$$

$$b(\vec{r},t) = \int d\vec{\kappa}\, T_b(\vec{\kappa}) A_\zeta(\vec{\kappa}) \exp\left[i(\vec{\kappa}\vec{r} - \Omega(\kappa)t)\right] + \text{c.c.} \qquad (\text{A.2})$$

where $T_a(\vec{\kappa})$ and $T_b(\vec{\kappa})$ T are respective transfer functions. Then

$$\left\langle a(\vec{r},t)b(\vec{r},t) \right\rangle$$

$$= \iint d\vec{\kappa}\, d\vec{\kappa}'\, T_a(\vec{\kappa}) T_b(\vec{\kappa}') \left\langle A_\zeta(\vec{\kappa}) A_\zeta(\vec{\kappa}') \right\rangle \exp\left\{ i\left[(\vec{\kappa} + \vec{\kappa}')\vec{r} - (\Omega(\kappa) + \Omega(\kappa')t) \right] \right\}$$

$$+ \iint d\vec{\kappa}\, d\vec{\kappa}'\, T_a^*(\vec{\kappa}) T_b(\vec{\kappa}') \left\langle A_\zeta^*(\vec{\kappa}) A_\zeta(\vec{\kappa}') \right\rangle \exp\left\{ i\left[(-\vec{\kappa} + \vec{\kappa}')\vec{r} + (\Omega(\kappa) - \Omega(\kappa')t) \right] \right\}$$

$$+ \iint d\vec{\kappa}\, d\vec{\kappa}'\, T_a(\vec{\kappa}) T_b^*(\vec{\kappa}') \left\langle A_\zeta(\vec{\kappa}) A_\zeta^*(\vec{\kappa}') \right\rangle \exp\left\{ i\left[(\vec{\kappa} - \vec{\kappa}')\vec{r} - (\Omega(\kappa) - \Omega(\kappa')t) \right] \right\}$$

$$+ \iint d\vec{\kappa}\, d\vec{\kappa}'\, T_a^*(\vec{\kappa}) T_b^*(\vec{\kappa}') \left\langle A_\zeta^*(\vec{\kappa}) A_\zeta^*(\vec{\kappa}') \right\rangle \exp\left\{ i\left[(-\vec{\kappa} - \vec{\kappa}')\vec{r} + (\Omega(\kappa) + \Omega(\kappa')t) \right] \right\} \qquad (\text{A.3})$$

Due to the assumed statistical homogeneity of roughness

$$\left\langle A_\zeta(\vec{\kappa}) A_\zeta^*(\vec{\kappa}') \right\rangle = \left\langle A_\zeta^*(\vec{\kappa}) A_\zeta(\vec{\kappa}') \right\rangle = \frac{1}{2} W_\zeta(\vec{\kappa}) \delta(\vec{\kappa} - \vec{\kappa}') \qquad (\text{A.4})$$

$$\left\langle A_\zeta(\vec{\kappa}) A_\zeta(\vec{\kappa}') \right\rangle = \left\langle A_\zeta^*(\vec{\kappa}) A_\zeta^*(\vec{\kappa}') \right\rangle = \frac{1}{2} W_\zeta(\vec{\kappa}) \delta(\vec{\kappa} + \vec{\kappa}') \qquad (\text{A.5})$$

which means

$$\left\langle a(\vec{r},t)b(\vec{r},t) \right\rangle = \frac{1}{2} \int d\vec{\kappa} \left[T_a(\vec{\kappa}) T_b^*(\vec{\kappa}) + T_a^*(\vec{\kappa}) T_b(\vec{\kappa}) \right] W_\zeta(\vec{\kappa})$$

$$+ \frac{1}{2} \int d\vec{\kappa}\, T_a(\vec{\kappa}) T_b(-\vec{\kappa}) W_\zeta(\vec{\kappa}) \left\langle \exp[-2i\Omega(\kappa)t] \right\rangle \qquad (\text{A.6})$$

$$+ \frac{1}{2} \int d\vec{\kappa}\, T_a^*(\vec{\kappa}) T_b^*(-\vec{\kappa}) W_\zeta(\vec{\kappa}) \left\langle \exp[2i\Omega(\kappa)t] \right\rangle$$

Reasonably assuming the ergodicity works here, and thus averaging over the total of the roughness realizations is equal to averaging over time, we write

$$\left\langle \cos\left[\Omega(\kappa)t\right]\right\rangle = \lim_{T\to\infty}\frac{1}{T}\int_{t}^{t+T}\cos\left[\Omega(\kappa)t'\right]dt' = 0 \qquad (A.7)$$

$$\left\langle \sin\left[\Omega(\kappa)t\right]\right\rangle = \lim_{T\to\infty}\frac{1}{T}\int_{t}^{t+T}\sin\left[\Omega(\kappa)t'\right]dt' = 0 \qquad (A.8)$$

Therefore

$$\left\langle \exp\left[2i\Omega(\kappa)t\right]\right\rangle = \left\langle \exp\left[-2i\Omega(\kappa)t\right]\right\rangle = 0 \qquad (A.9)$$

and thus

$$\left\langle a(\vec{r},t)b(\vec{r},t)\right\rangle = \int d\vec{\kappa}\,\mathrm{Re}\left[T_a(\vec{\kappa})T_b^*(\vec{\kappa})\right]W_\zeta(\vec{\kappa}) \qquad (A.10)$$

We set $a = \delta\sigma_0^{\pm}/\langle\sigma_0^{\pm}\rangle$ and $b = v_{\mathrm{rad}}$, and find

$$r_{v,\delta\sigma_0}^{\pm} = \frac{\langle v_{\mathrm{rad}}\delta\sigma_0^{\pm}\rangle}{\sigma_{\mathrm{rad}}\sigma_{\delta\sigma_0}^{\pm}} = \frac{\langle\sigma_0^{\pm}\rangle}{\sigma_{\mathrm{rad}}\sigma_{\delta\sigma_0}^{\pm}}\int d\vec{\kappa}\,\mathrm{Re}\left[T(\vec{\kappa})T_{v_{\mathrm{rad}}}^*(\vec{\kappa})\right]W_\zeta(\vec{\kappa}) \qquad (A.11)$$

Appendix B

We locate the ground range x value and analyse an image line along the Y-axis. For simplicity, we assume that the roughness is propagating azimuthally; consequently, the $v_{rad}(y')$ process can be regarded as one-dimensional, therefore we shall further work with the full dv_{rad}/dy' and not with a partial derivative in Appendix B. According to Tikhonov (1986), the average number $\langle N \rangle$ of intersections between the random process $(R/V)\widehat{v}_{rad}(y')$ and the straight line $y - y'$ or, which is the same, the average number of zero crossing of the random process $\widehat{v}_{rad}(y') - (V/R)(y - y')$ is given by

$$\langle N \rangle = \iint d\xi d\eta |\eta| p\left(\frac{V}{R}\xi, \eta - \frac{V}{R}\right) \tag{B.1}$$

where $\xi = y - y'$, $\eta = \left(d\widehat{v}_{rad}/dy'\right) + (V/R)$ and $p(\cdot)$ denotes the joint PDF of the process and its derivative.

Since the joint PDF of Gaussian process (which we assume is characteristic of $\widehat{v}_{rad}$) is

$$p\left(\widehat{v}_{rad}, \frac{d\widehat{v}_{rad}}{dy'}\right) = \frac{1}{2\pi\widehat{\sigma}_{rad}[-B''_{rad}(0)]^{1/2}} \exp\left\{-\frac{1}{2}\left[\frac{\widehat{v}^2_{rad}}{\widehat{\sigma}^2_{rad}} + \frac{\left(d\widehat{v}_{rad}/dy'\right)^2}{-B''_{rad}(0)}\right]\right\} \tag{B.2}$$

where B''_{rad} is the second derivative of the correlation function of $\widehat{v}_{rad}$; relation (B.1) reduces to

$$\langle N \rangle = \frac{1}{2\pi\widehat{\sigma}_{rad}[-B''_{rad}(0)]^{1/2}} \iint d\xi \, d\eta |\eta| \times \exp\left\{-\frac{1}{2}\left[\frac{V^2}{R^2\widehat{\sigma}^2_{rad}}\xi2 + \frac{(\eta - V/R)^2}{-B''_{rad}(0)}\right]\right\} \tag{B.3}$$

This integral can be found analytically. Integrating consecutively first over ξ, and then over η, we find

$$\langle N \rangle = \frac{2}{\sqrt{2\pi}} \frac{R}{V} \left[-B''_v(0)\right]^{1/2} \exp\left\{-\frac{V^2}{2R^2[-B''_{rad}(0)]}\right\} + \mathrm{erf}\left\{\frac{V}{\sqrt{2}R[-B''_{rad}(0)]^{1/2}}\right\} \tag{B.4}$$

where

$$\mathrm{erf}(x) = \frac{2}{\sqrt{\pi}} \int_0^x e^{-t^2} dt \tag{B.5}$$

The characteristic wavelength along the Y-axis is defined as

$$\Lambda_y = \frac{2\pi}{\langle \kappa^2 \rangle^{1/2}} \tag{B.6}$$

In turn,

$$\langle \kappa^2 \rangle = \frac{1}{\sigma_{rad}^2} \int d\kappa \, \kappa^2 \hat{W}_{rad}(\kappa) \tag{B.7}$$

where

$$\hat{W}_{rad}(\kappa) = \frac{1}{2\pi} \int d\rho B_{rad}(\rho) \, \exp(-i\kappa\rho) \tag{B.8}$$

Consequently,

$$B_{rad}(\rho) = \int d\kappa \hat{W}_{rad}(\kappa) \, \exp(i\kappa\rho) \tag{B.9}$$

which gives

$$-B_{rad}''(0) = \int d\kappa \, \kappa^2 \hat{W}_{rad}(\kappa) = \sigma_{rad}^2 \langle \kappa^2 \rangle \tag{B.10}$$

Therefore

$$\left[-B_{rad}''(0) \right]^{1/2} = \frac{2\pi}{\Lambda_y} \sigma_{rad} \tag{B.11}$$

Having substituted Eqn. (B.11) into (B.4) and taking into account that $\beta_y = (R/V)(\sigma_{rad}/\Lambda_y)$ we obtain

$$\langle N \rangle = 2\sqrt{2\pi}\beta_y \exp\left[-\left(2\sqrt{2\pi}\beta_y \right)^{-2} \right] + \mathrm{erf}\left[\left(2\sqrt{2\pi}\beta_y \right)^{-1} \right] \tag{B.12}$$

Appendix C

We find the average value of the denominator in Eqns (5.45) and (5.46) also assuming that $\widehat{v}_{\mathrm{rad}}$ has a Gaussian distribution. As the value is connected with $\widehat{v}_{\mathrm{rad}}$ linearly, the derivative over the azimuthal coordinate $\partial \widehat{v}_{\mathrm{rad}}/\partial y'$ is also characterized by the Gaussian distribution with zero mean, and for its RMS we shall take

$$\frac{2\pi \widehat{\sigma}_{\mathrm{rad}}}{\Lambda_v} = 2\pi \beta_v \frac{V}{R} \tag{C.1}$$

Then, having introduced a new variable $u = (R/V)(\partial \widehat{v}_{\mathrm{rad}}/\partial y')$ we can write

$$\left\langle \left| 1 + \frac{R}{V} \frac{\partial \widehat{v}_{\mathrm{rad}}}{\partial y'} \right| \right\rangle = \langle |1 + u| \rangle$$

$$= \frac{1}{(2\pi)^{3/2} \beta_v} \left[\int_{-1}^{\infty} (1 + u) \exp\left(-\frac{u^2}{8\pi^2 \beta_v^2} \right) du \right.$$

$$\left. - \int_{-\infty}^{-1} (1 + u) \exp\left(-\frac{u^2}{8\pi^2 \beta_v^2} \right) du \right] \tag{C.2}$$

$$= \frac{2}{(2\pi)^{3/2} \beta_v} \left[\int_0^1 \exp\left(-\frac{u^2}{8\pi^2 \beta_v^2} \right) du + \int_0^{\infty} u \exp\left(-\frac{u^2}{8\pi^2 \beta_v^2} \right) du \right.$$

$$\left. - \int_0^1 u \exp\left(-\frac{u^2}{8\pi^2 \beta_v^2} \right) du \right]$$

These integrals are easily obtained analytically, and we get

$$\left\langle \left| 1 + \frac{R}{V} \frac{\partial \widehat{v}_{\mathrm{rad}}}{\partial y'} \right| \right\rangle = \langle N \rangle \tag{C.3}$$

References

Alpers, W.R. 1983. Monte Carlo simulations for studying the relationship between ocean wave and synthetic aperture radar spectra. J. Geophys. Res., vol. 88, no. C3, pp. 1745–1759.

Alpers, W.R. and C. Brüning 1986. On the relative importance of notion–related contributions to the SAR imaging mechanism of ocean surface waves. IEEE Trans. Geosci. Remote Sens., vol. GE-24, no. 6, pp. 873–885.

Alpers, W.R. and K. Hasselmann 1978. The two-frequency microwave technique for measuring ocean wave spectra from an airplane of satellite. Boundary Layer Meteorol., vol. 13, pp. 215– 230.

Alpers, W.R., D.B. Ross, and C.L. Rufenach 1981. On the detectability of ocean surface waves by real and synthetic aperture radar. J. Geophys. Res., vol. 86, no. C7, pp. 6481–6498.

Alpers, W. and K. Hasselmann 1982. Spectral signal to clutter and thermal noise properties of ocean wave imaging synthetic aperture radars. Int. J. Remote Sens., vol. 3, no. 4, pp. 423–446.

Alpers, W.R. and C.L. Rufenach 1979. The effect of orbital motions on synthetic aperture radar imagery of ocean waves. IEEE Trans. Antennas Propag., vol. 27, no. 5, pp. 685–690.

Apel, J.R. 1994. An improved model of the ocean surface wave vector spectrum and its effects on radar backscatter. J. Geophys. Res., vol. 99, pp. 16269–16291.

Barrick, D.E. 1974. Wind dependence of quasi-specular microwave sea scatter. IEEE Trans. Antennas Propag., vol. AP-22, pp. 135–136.

Barrick, D.E. 1978. HF radio oceanography – A review. Boundary Layer Meteorol., vol. 13, pp. 23–44.

Barrick, D.E., M.W. Ewans, and V.L. Weber 1977. Ocean surface currents mapped by radar. Science, vol. 198, pp. 138–198.

Basovich, A.Ya., V.V. Bakhanov, D.M. Bravo-Zhivotovsky, L.S. Dolin, Yu.M. Zhidko, A.A. Zaitsev, N.A. Zavol'sky, E.M. Zuikova, B.F. Kelbalikhanov, A.G. Luchinin, V.V. Narozhny, V.I. Titov, and A.B. Shmagin 1985. Impact short trains of internal waves on the sea wind roughness (in Russian). Dokl. AN SSSR, vol. 283, no. 1, pp. 209–212.

Bass, F.G., I.M. Fuks, A.I. Kalmykov, I.E. Ostrovsky, and A.D. Rozenberg 1968. Very high frequency radio wave scattering by a disturbed sea surface. IEEE Trans. Antennas Propag., vol. AP-16, no. 2, pp. 554–568.

Bass, F.G. and I.M. Fuks 1979. Wave Scattering from Statistically Rough Surfaces, Oxford: Pergamon Press, 525 pp.

Beal, R.C., P.S. De-Leonibus, and I. Katz (eds.) 1981. Spaceborne Synthetic Aperture Radar for Oceanography, Baltimore: John Hopkins University Press, 215 pp.

Boerner, W.-M. 2007. Applications of polarimetric and interferometric SAR to environmental remote sensing and its activities: recent advances in extrawide band polarimetry, interferometry and polarimetric interferometry in synthetic aperture remote sensing and its applications. In Radar Polarimetry and Interferometry, pp. 5-1–5-34. Edicational Notes RTO-EN-SET-081bis. Available from: http://www.rto.nato.int/abstracts.asp

Braude, S.Ya. (ed.) 1962. Radio-Oceanographic Investigations of the Sea Roughness (in Russian), Kiev: Publication of Ukraine Academy of Sciences.

Brekhovskikh, L.M. 1952. The diffraction of waves by a rough surface (in Russian). Zh. Exper. i Teor. Fiz., vol. 23, no. 3, pp. 275–288.

Brüning, C., W. Alpers, and K. Hasselmann 1990. Monte Carlo simulation studies of the nonlinear imaging of a two dimensional surface wave field by a synthetic aperture radar. Int. J. Remote Sens., vol. 11, no. 10, pp. 1695–1721.

Bulatov, M.G., Yu.A. Kravtsov, O.Yu. Lavrova, K.Ts. Litovchenko, M.I. Mityagina, M.D. Raev, K.D. Sabinin, Yu.G. Trokhimovskii, A.N. Churyumov, and I.V. Shugan 2004. Physical mechanisms of aerospace radar imaging of the ocean (in Russian). Usp. Fiz. Nauk, vol. 173, no. 1, pp. 69–87.

Carande, R.E. 1992. Dual baseline and frequency along-track interferometry. Proc. IGARSS'92, pp. 1585–1588.

Carlson, H., K. Richter, and H. Walden 1967. Messungen der statistichen verteilung der auslenkung der meeresoberflasche im seegang. Dtsch. Hydrogr. Z., vol. 20, no. 2, pp. 59–64.

Crombie, D.D. 1955. Doppler spectrum of sea echo at 13.56 Ms/s. Nature, vol. 175, p. 681

Crombie, D.D., K. Hasselmann, and W. Sell 1978. High-frequency radar observations os sea waves traveling in opposition to the wind. Boundary Layer Meteorol., vol. 13, pp. 45–54.

Davidan, I.N., L.I. Lopatoukhin, and V.A. Rozhkov 1985. Wind Waves in the World Ocean (in Russian), Leningrad, Gidrometeoizdat, 256 pp.

Dean, R.G. and R.A. Dalrymple 1991. Water wave mechanics for engineers and scientists. Advanced Series on Ocean Engineering, vol. 2. Singapore: World Scientific, 353 pp.

Elachi, Ch., T. Bicknell, R.L. Jordan, and Ch. Wu 1982. Spaceborne synthetic aperture imaging radars: Applications, techniques, and technology. Proc. IEEE, vol. 70, no. 10, pp. 1174–1209.

Elfouhaily, T., B. Chapron, and K. Katsaros 1997. A unified directional spectrum for long and short wind-driven waves. J. Geophys. Res., vol. 102, no. C7, pp. 15781–15796.

Elfouhaily, T., D.R. Tompson, D.E. Freund, D. Vandemark, and B. Chapron 2001. A new bistatic model for electromagnetic scattering from perfectly-conducting fandom surfaces: Numerical evaluation and comparison with SPM. Waves Random Media, vol. 9, pp. 281–294.

Elfouhaily, T., D.R. Tompson, D. Vandemark, and B. Chapron 1999. A new bistatic model for electromagnetic scattering from perfectly-conducting fandom surfaces. Waves Random Media, vol. 11, pp. 33–43.

Elfouhaily, T.M. and J.T. Johnson 2007. A new model for rough surface scattering. IEEE Trans. Geosci. Remote Sens., vol. 45, no. 7, pp. 2300–2308.

Engen, G. and H. Johnsen 1995. SAR-Ocean wave inversion using image cross spectra. IEEE Trans. Geosci. Remote Sens., vol. GE-33, no. 4, pp. 1047–1056.

Engen, G., P.W. Vachon, H. Johnsen, and F.W. Dobson 2000. Retrieval of ocean surface wave spectra and RAR MTFs from dual-polarization SAR data. IEEE Trans. Geosci. Remote Sens., vol. 38, pp. 391–403.

Farquharson, G., W.N. Junek, A. Ramanathan, S.I. Frasier, R. Tessier, D.J. McLaughlin, M.A. Sletten, and J.V. Toporkov 2004. A pod-based dual-beam SAR. IEEE Geosci. Remote Sens. Lett., vol. 1, no. 2, pp. 62–65.

Frasier, S.J. and A.J. Camps 2001. Dual-beam Interferometry for ocean surface current vector mapping. IEEE Trans. Geosci. Remote Sens., vol. 39, no. 2, pp. 403–414.

Fuks, I.M. 1966. On the theory of radio wave scattering by a disturbed sea surface (in Russian). Izv.VUZ – Radiofizika, vol. 9, no. 5, pp. 876–887.

Fuks, I.M. 1974. Spectrum width of the signals scattered by the rough sea surface (in Russian). Akust. Zh., vol. 20, no. 3, pp. 458–468.

Fuks, I. and V. Zavorotny 2007. Polarization dependence of radar contrast for sea surface oil slicks. Proceedings of IEEE Radar Conference, 17–20 April, Boston, MA, USA, pp. 503–508.

Fung, A.K. 1994. Microwave Scattering and Emission Models and Their Applications, Boston MA: Artech House, 573 pp.

Fung, A.K. and K.K. Lee 1982. A semi-empirical sea-spectrum model for scattering coefficient estimation. IEEE J. Oceanic Eng., vol. OE-7, no. 4, pp. 166–176.

Furevik, B.R. and E. Korsbakken 2000. Comparison of derived wind speed from synthetic aperture radar and scatterometer during the ERS tandem phase. IEEE Trans Geosci. Remote Sens., vol. GE-38, no. 2, pp. 1113–1121.

Gade, M., W. Alpers, H. Huefnerfuss, H. Masuko, and T. Kobayashi 1998. Imaging of biogenic and antropogenic ocean surface films by the multifrequency/multipolarization SIR-C/X-SAR. J. Geophys. Res., vol. 103, no. C9, pp. 18851–18866.

Gerling, T.W. 1986. Structure of the surface wind field from the Seaset SAR. J. Geophys. Res., vol. 91, pp. 2308–2320.

Goldstein, R.M., T.P. Barnett, and H.A. Zebker 1989. Remote sensing of ocean currents. Science, vol. 246, pp. 1282–1285.

Goldstein, R.M. and H.A. Zebker 1987. Interferometric radar measurement of ocean surface currents. Nature, vol. 328, pp. 707–709.

Grebenyuk, Yu.V., M.B. Kanevsky, and V.Yu. Karaev 1994. Doppler spectrum width of microwave radar signal backscattered from the sea surface at moderate incidence angles (in Russian). Izv. AN SSSR – FAO, vol. 30, no. 1, pp. 59–62.

Hara, T. and J.W. Plant 1994. Hydrodynamic modulation of short wind-wave spectra by long waves and its measurement using microwave backscatter. J. Geophys. Res., vol. 99, no. C9, pp. 9767–9784.

Hasselmann, K., T.P. Barnett, E. Bouws, H. Carlson, D.E. Cartright, K. Enke, J.A. Ewing, H. Gienapp, D.E. Hasselmann, P. Kruseman, A. Meerburg, P. Muller, D.J. Oibers, K. Richter, W. Sell, and H. Walden 1973. Measurements of wind-wave growth and swell decay during the Joint North Sea Wave Project (JONSWAP). Deut. Hydrogr. Z. Reihe A, no. 12, 1–95.

Hasselmann, K. and S. Hasselmann 1991. On the nonlinear mapping of an ocean wave spectrum into a synthetic aperture radar image spectrum and its inversion. J. Geophys. Res., vol. 96, no. C6, pp. 10713–10729.

Hasselmann, K, R.K. Raney, W.J. Plant, W. Alpers et al. 1985. Theory of SAR ocean wave imaging: The MARSEN view. J. Geophys. Res., vol. 90, no. C3, pp. 4659–4686.

Hauser, D. and G. Caudal 1996. Combined analysis of the radar cross-section modulation due to the long ocean waves around 14° and 34° incidence: Implication for the hydrodynamic modulation. J. Geophys. Res., vol. 101, no. C11, pp. 25833–25846.

He, Y., W. Perrie, T. Xie, and Q. Zou 2004. Ocean wave spectra from a linear polarimetric SAR. IEEE Trans. Geosci. Remote Sens., vol. 42, no. 11, pp. 2623–2631.

Holt, B. 2004. SAR imaging of the ocean surface. In Synthetic Aperture Radar Marine User's Manual (ed. by C.R. Jackson and J.R. Apel), Washington, DC: US Department of Commerce, NOAA, September, pp. 25–80.

Horstmann, J., H. Schiller, J. Schulz-Stellenfleth, and S. Lehner 2003. Global wind speed retrieval from SAR. IEEE Trans. Geosci. Remote Sens., vol. 41, no. 10, pp. 2277–2286.

Ivonin, D.V., P. Broche, J.-L. Devenon, and V.I. Shrira 2004. Validation of HF radar probing of the vertical shear of surface currents by acoustic Doppler current profiler measurements. J. Geophys. Res., vol. 109, C040003, doi:10.1029/2003JC002025, 2004.

Johannessen J.A. 1995. The potential in using synthetic aperture radar for coastal environmental monitoring: The ERS-1 experience. Proceedings of Colloquium "Operation oceanography and satellite observation", 16–20 October, Biarritz, France.

Kalmykov, A.I. and V.V. Pustovoitenko 1976. On polarization features of radio signals scattered from sea surface at small grazing angles. J. Geophys. Res., vol 51, pp. 1960–1964.

Kanevsky, M.B. 1982. Formation of radar image of internal waves on the sea surface (in Russian). In Impact Large-cale Internal Waves on the Sea Surface (ed. by E.N. Pelinovsky), Gorky: Institute of Applied Physics, 251 pp.

Kanevsky, M.B. 1985. A spectrum of radar image of the sea surface (in Russian). Izv. AN SSSR – FAO, vol. 21, no. 5, pp. 544–550.

Kanevsky, M.B. 1993. On the theory of SAR ocean wave imaging. IEEE Trans. Geosci. Remote Sens., vol. 31, no. 5, pp. 1031–1035.

Kanevsky, M.B. 2002. A new technique for estimating the spectrum of SAR images of the ocean surface: Theory and experimental verification. Proceedings of EUSAR 2002, Cologne, Germany, pp. 737–740.

Kanevsky, M.B. 2005. New spectral estimate for SAR imaging of the ocean. Int. J. Remote Sens., vol. 26, no. 17, pp. 3707–3715.

Kanevsky, M.B., S.A. Ermakov, E.M. Zuikova, V.Yu. Karaev, V.Yu. Goldblat, L.A. Sergievskaya, Yu.B. Shchegol'kov, J.C. Scott, and N. Stapleton 1997. Experimental investigation of Doppler spectra of microwave signals backscattered by sea slicks. Proc. IGARSS'1997, vol. IV, pp. 1530–1532.

Kanevsky, M.B. and V.Yu. Karaev 1996a. Spectral characteristics of superhigh frequency (SHF) radar signal backscattered by the sea surface at small incidence angles (in Russian). Izv. VUZ – Radiofizika, vol. 39, no. 5, pp. 517–526.

Kanevsky, M.B. and V.Yu. Karaev 1996b. The microwave radar signal Doppler spectrum and the problem of ocean surface slick detection. Proc. IGARSS'96, vol. III, pp. 1493–1495.

Kanevsky, M.B., V.Yu. Karaev, L.V. Lubyako, E.M. Zuikova, V.Yu. Goldblat, V.I. Titov, and G.N. Balandina 2001. Doppler spectra of centimeter and millimeter microwaves backscattered from a rough water surface. Radiophys. Quantum Electron., vol. 44, no. 11, pp. 850–857.

Kanevsky, M.B and L.V. Novikov 1990. On the theory of SAR imaging of azimuthally traveling sea waves (in Russian). Issled Zemli iz Kosmosa, no. 4, pp. 12–18.

Karaev, V.Yu. and G.N. Balandina 2000. Modified wave spectrum and remote sensing of the ocean (in Russian). Issled Zemli iz Kosmosa, no. 5, pp. 1–12.

Karaev, V.Yu., M.B. Kanevsky, G.N. Balandina, P. Challenor, C. Gommenginger, and M. Srokosz 2005. The concept of microwave radar with an asymmetric knife-like beam for the remote sensing of ocean waves. J. Atmos. Oceanic Technol., vol. 22, pp. 1809–1820.

Karaev, V.Yu., M.B. Kanevsky, G.N. Balandina, P.D. Cotton, P.G. Challenor, C.P. Gommenginger, and M.A. Srokosz 2002. On the problem of the near ocean surface wind speed retrieval by radar-altimeter: A two-parameter algorithm. Int. J. Remote Sens., vol. 23, no. 16, pp. 3263–3283.

Karaev, V.Yu., M.B. Kanevsky, G.N. Balandina, and C. Gommenginger 2002b. On the regional features influence on the accuracy of a near surface wind speed retrieval over ocean as applied to altimeter measurements (in Russian). Issled. Zemli iz Kosmosa, no. 4, pp. 32–42.

Karaev V.Yu., M.B. Kanevsky, G.N. Balandina, E.M. Meshkov, P. Challenor, M. Srokosz, and C. Gommenginger 2006b. A rotating knife-like altimeter for wide-swath remote sensing of the ocean: Wind and waves. Sensors, no. 6, pp. 260–281. www.mdpi.org/sensors/list06.htm#new

Karaev, V.Yu., M.B. Kanevsky, E.M. Meshkov, D. Cotton, and C. Gommenginger 2006a. Retrieval of the near-surface wind speed from satellite altimeter data: A review of algorithms. Radiophys. Quantum Electron., vol. 49, no. 4, pp. 279–293.

Kasilingam, D.P. and O.H. Shemdin 1990. Models for synthetic aperture radar imaging of the ocean: A comparison. J. Geophys. Res., vol. 95, no. C9, pp. 16263–16276.

Keyte, G.E. and J.T. Macklin 1986. SIR-B observations of ocean waves in NE Atlantic. IEEE Trans. Geosci. Remote Sens., vol. GE-24, no. 4, pp. 552–558.

Kim Duk-jin, W.M. Moon, D. Moller, and D.A. Imel 2003. Measurements of ocean surface waves and currents using L- and C-band along-track interferometric SAR. IEEE Trans. Geosci. Remote Sens., vol. 41, no. 12, pp. 2821–2832.

Kitaigorodskii, S.A. 1973. The Physics of Air-Sea. Jerusalem: Keter Press, pp. 9–73.

Krogstad, H.E, O. Samset, and P.W. Vachon 1994. Generalization of the nonlinear ocean SAR transform and a simplified SAR inversion algorithm. Atmosphere-Ocean, vol. 32, pp. 61–82.

Kropfli, R.A. and S.F. Clifford 1994. The San Clemente ocean probing experiment: A study of air-sea interactions with remote and in situ sensors. Proceedings of IGARSS'94, Pasadena, CA, pp. 2407–2409.

Kuryanov, B.F. 1962. Sound scattering on a rough two-scale surface (in Russian). Akust. Zh., vol. 8, no. 3, pp. 325–333.

Lee, J.S., D.L. Shuler, and T.L. Ainsworth 2000. Polarimetric SAR data compensation for terrain azimuth slope variation. IEEE Trans. Geosci. Remote Sens., vol. 38, pp. 2153–2163.

Lee, P.H.Y., J.D. Barter, K.L. Beach, E. Caponi, C.L. Hindman, B.M. Lake, H. Rungaldier, and J.C. Shelton 1995b. Power spectral lineshapes of microwave radiation backscattered from sea surfaces at small grazing angles. IEE Proc. – Radar, Sonar, Navigation, vol. 142, no. 5, pp. 252–258.

Lee, P.H.Y., J.D. Barter, K.L. Beach, C.L. Hindman, B.M. Lake, H. Rungaldier, J.C. Shelton, A.B. Williams, R. Yee, and H.C. Yuen 1995a. X-band microwave backscattering from ocean waves. J. Geophys. Res., vol. 100, no. C2, pp. 2591–2611.

Lee, P.H.Y., J.D. Barter, K.L. Beach, B.M. Lake, H. Rungaldier, H.R. Thompson, Jr., and R. Yee 1998. Scattering from breaking waves without wind. IEEE. Trans. Antennas Propag., vol. 46, no. 1, pp. 14–25.

Lee, P.H.Y., J.D. Barter, E. Caponi, M. Caponi, C.L. Hindman, B.M. Lake, and H. Rungaldier 1996. Wind-speed dependence of small-grazing angle microwave backscatter from sea surfaces. IEEE Trans. Antennas Propag., vol. 44, no. 3, pp. 333–340.

Lementa, Yu.A. 1980. Scattering of UHF radio waves from the rough sea surface and determination of the sea state parameters trough backscatter characteristics. Candidate of sciences Thesis (in Russian), Institute of Radiophysics and Electronics, Kharkov, USSR.

Levin, B.R. 1969. Statistical Radio Engineering, vol. I (in Russian) Sov. Radio, 751 pp.

Li, F., D. Held, and J. Curlander 1982. Doppler parameter estimation techniques for spaceborn SAR with application to ocean current measurement. Proceedings of IGARSS'82, pp. 7.1–7.5.

Longuet-Higgins, M.S. 1952. On the statistical distribution of the height of sea waves. J. Mar. Res., vol. 11, no. 3, pp. 245–266.

Longuet-Higgins, M.S., D.E. Cartwright, and N.D. Smith 1963. Observations of the directional spectrum of sea waves using the motions of a floating buoy. In Ocean Wave Spectra, Englewood Cliffs, NJ: Prentice-Hall, pp. 111–136.

Lui, P.L.F, S.B. Yoon, and J.T. Kirby 1985. Non-linear refraction-diffraction of waves in shallow water. J. Fluid Mech., vol. 153, pp. 185–201.

Lyzenga, D.R. 1991. Interaction of short surface and electromagnetic waves with ocean fronts. J. Geophys. Res., vol. 96, no. C6, pp. 10765–10772.

Lyzenga, D.R. 2002. Unconstrained inversion of wave height spectra. IEEE Trans. Geosci. Remote Sens., vol. 40, pp. 261–270.

Lyzenga, D.R., A.L. Maffett, and R.A. Shuchman 1983. The contribution wedge scattering to the radar cross section of the ocean surface. IEEE Trans. Geosci. Remote Sens., vol. GE-21, no. 4, pp. 502–505.

Lyzenga, D.R. and G.O. Marmorino 1998. Measurement of surface currents using sequential synthetic aperture radar images of slicks patterns near the edge of the Gulf Stream. J. Geophys. Res., vol. 103, no. C9, pp. 18769–18777.

Marple, S.L., Jr. 1987. Digital Spectral Analysis with Applications, Prentice-Hall Inc., Englewood Cliffs, NJ.

Mastenbroek, C. and C.F. de Valk 2000. A semiparametric algorithm to retrieve ocean wave spectra from synthetic aperture radar. J. Geophys. Res., vol. 105, no. C2, pp. 3497–3516.

Melief, H.W., H. Greidanus, P. van Genderen, and P. Hoogeboom 2006. Analysis of sea spikes in radar sea clutter data. IEEE Trans. Geosci. Remote Sens., vol. 44, no. 4, pp. 985–993.

Mercier, G. and F. Girard-Ardhuin 2005. Partially supervised oil-slick detection by SAR imagery using kernel expansion. IEEE Trans. Geosci. Remote Sens., vol. 44, no. 10, pp. 2839–2846.

Miller, A.R., R.M. Brown, and E. Veigh 1984. New derivation for the rough-surface reflection coefficient and for the distribution of sea-wave elevation. IEE Proc., Pt. H, vol. 131, no. 2, pp. 114–116.

Miller, A.R. and E. Veigh 1986. Family of curves for the rough-surface reflection coefficient. IEE Proc., Pt. H, vol. 133, no. 6, pp. 483–489.

Mitsuyasu, H., F. Tasai, T. Suhara, S. Misuno, M. Ohkuso, T. Honda, and K. Rindiishi 1975. Observations of the directional spectrum of ocean waves using a cloverleaf buoy. J. Phys. Oceanogr., vol. 5, pp. 750–760.

Monaldo, F.M., and R. Beal 2004. Wind speed and direction. In Synthetic Aperture Radar Marine User's Manual (ed. by C.R. Jackson and J.R. Apel), Washington, DC: US Department of Commerce, NOAA, September, pp. 305–320.

Ohlmann, C., P. White, L. Washburn, E. Terrill, B. Emery, and M. Otero 2007. Interpretation of coastal HF radar-derived surface currents with high-resolution drifter data. J. Atmos. Ocean. Tech., vol. 24, doi:10.1175/JTECH1998.1.

Ouchi, K. and H. Wang 2005. Interlook cross-correlation function of speckle in SAR images of sea surface processed with partially overlapped subapertures. IEEE Trans. Geosci. Remote Sens., vol. 43, no. 4, pp. 695–701.

Pereslegin, S.V. 1975a. Connection between microwave scattering and spatio-temporal characteristics of the developed sea roughness (in Russian). Izv. AN SSSR, ser, FAO, vol. 11, no. 5, pp. 481–490.

Pereslegin, S.V. 1975b. Statistical characteristics of microwave scattering from the sea surface accounting a finite resolution and the depolarization factor (in Russian). Izv. AN SSSR, ser, FAO, vol. 11, no. 6, pp. 610–618.

Phillips, O.M. 1985. Spectral and statistical properties of the equilibrium range in the wind-generated gravity waves. J. Fluid Mech., vol. 156, pp. 505–531.

Phillips, O.M. 1988. Radar returns from the sea surface – Bragg scattering and breaking waves. J. Phys. Oceanogr., vol. 18, pp. 1065–1074.

Pichel, W.G., P. Clemente-Colon, C.C. Wackerman, and K.S. Friedman 2004. Ship and wake detection. In Synthetic Aperture Radar Marine User's Manual (ed. by C.R. Jackson and J.R. Apel), Washington, DC: US Department of Commerce, NOAA, September, pp. 277–304.

Pidgeon, V.W. 1968. Doppler dependence of radar sea return. J. Geophys. Res., vol. 73, pp. 1333–1341.

Pierson, W.J. and Z.A. Moskovtz 1964. A proposed spectral form for fully developed wind seas based on the similarity theory of S.A. Kitaigorodskii. J. Oceanogr. Res., vol. 69, pp. 5181–5190.

Pierson, W.J. and R.A. Stacy 1973. The elevation, slope, and curvature spectra of wind roughened surface. NASA Rep., CR-2247, 129 pp.

Plant, W.J. and W.C. Keller 1990. Evidence of Bragg scattering in microwave Doppler spectra of sea return. J. Geophys. Res., vol. 95, no. C9, pp. 16299–16310.

Plant, W.J., W.C. Keller, and A. Cross 1983. Parametric dependence of ocean wave – radar modulation transfer functions. J. Geophys. Res., vol. 88, no. C14, pp. 9747–9756.

Poulter, E.M., M.J. Smith, and J.A. McGregor 1994. Microwave backscatter from the sea surface: Bragg scattering by short gravity waves. J. Geophys. Res., vol. 99, no. C4, pp. 7929–7943.

Rodriguez, E., D. Imel, and B. Houshmand 1995. Two-dimensional surface currents using vector along-track Interferometry. Proceedings of PIERS'95, Seattle, WA, p. 763.

Romeiser, R., H. Breit, M. Eineder, H. Runge, P. Flament, K. de Jong, and J. Vogelzang 2005. Current measurements by SAR along-track interferometry from a space shuttle. IEEE Trans. Geosci. Remote Sens., vol. 43, no. 10, pp. 2315–2324.

Romeiser, R and D.R. Thompson 2000. Numerical study on the along-track interferometric radar imaging mechanism of oceanic surface currents. IEEE Trans. Geosci. Rem. Sens., vol. 38, no. 1, pp. 446–458.

Rozenberg A.D. 1980. Investigation of the sea surface by radio and acoustic methods (in Russian). Doctor of Sciences Thesis, Institute of Oceanology AN SSSR, Moscow.

Rozenberg, A.D., I.E. Ostrovsky, and A.I. Kalmykov 1966. Frequency shift at scattering of radio waves by the rough sea surface (in Russian). Izv. VUZ-Radiofizika, vol. 9, no. 2, pp. 233–240.

Rozenberg, A.D., I.E. Ostrovsky, V.I. Zeldis, I.A. Leykin, and V.G. Ruskevich 1973. Determination of energetic part of the sea roughness spectrum using phase characteristics of backscattered signals (in Russian). Izv. AN SSSR ser. FAO, vol. 9, no. 12, pp. 1323–1326.

Rufenach, C.L., R.B. Olsen, R.A. Shuchman, and C.A. Russel 1991. Comparison of aircraft synthetic aperture radar and buoy spectra during the Norwegian continental shelf experiment. J. Geophys. Res., vol. 96, no. C6, pp. 10423–10441.

Rytov, S.M., Yu.A. Kravtsov, and V.I. Tatarskii 1989. Principles of Statistical Radio Physics. Vol. 4: Wave Propagation Through the Random Media, Berlin, Heidelberg, Springer Verlag.

Schmidt, A., C. Brüning, and W. Alpers 1993. Ocean wave – modulation transfer functions inferred from airborne synthetic aperture radar imagery. Proceedings of OCEANS'93, Canada, Victoria, 18–21 October, pp. III-310–III-315.

Schmidt, A., V. Wismann, R. Romeiser, and W. Alpers 1995. Simultaneous measurements of the ocean wave-radar modulation transfer function at L,C, and X bands from the research platform *Nordsee*. J. Geophys. Res., vol. 100, no. C5, pp. 8815–8827.

Schuler, D.L., R.W. Jansen, J.S. Lee, and D. Kasilingam 2003. Polarization orientation angle measurements of ocean internal waves and current fronts using polarimetric SAR. IEE Proc. Radar, Sonar Navigation, vol. 150, no. 3, pp. 135–143.

Schuler, D.L. and J.S. Lee 1995. A microwave technique to improve the measurement of directional ocean wave spectra. Int. J. Remote Sens., vol. 16, no. 2, pp. 199–215.

Schuler, D.L., J.S. Lee, D. Kasilingam, and E. Potter 2004. Measurement of ocean surface slopes and wave spectra using polarimetric SAR data. Remote Sens. Environ., vol. 91, pp. 198–211.

Schulz-Stellenfleth, J., J. Horstmann, S. Lehner, and W. Rosenthal 2001. Sea surface imaging with an across-track interferometric synthetic aperture radar: The SINEWAVE experiment. IEEE Trans. Geosci. Remote Sens., vol. 39, no. 9, pp. 2017–2028.

Schulz-Stellenfleth, J. and S. Lehner 2001. Ocean wave imaging using an airborne single pass across-track interferometric SAR. IEEE Trans. Geosci. Remote Sens., vol. 39, no. 1, pp. 38–45.

Schulz-Stellenfleth, J. and S. Lehner 2004. Measurement of 2-D sea surface elevation fields using complex synthetic aperture radar data. IEEE Trans. Geosci. Remote Sens., vol. 42, no. 6, pp. 1149–1160.

Shay, L.K., J. Martinez-Pedraja, T.M. Cook, and B.K. Haus 2007. High-frequency radar mapping of surface currents using WERA. J. Atmos. Ocean. Tech., vol. 24, doi:10.1175/JTECH1985.1.

Shrira, V.I., D.V. Ivonin, P. Broche, and J.C. de Maistre 2001. On remote sensing of vertical shear of ocean surface currents by means of a single frequency VHF radar. Geophys. Res. Lett., vol. 28, pp. 3955–3958.

Sletten, M.A. 1998. Multipath scattering in ultrawide-band radar sea spikes. IEEE. Trans. Antennas Propag., vol. 46, no. 1, pp. 45–56.

Sletten, M.A., J.C. West, X. Liu, and J.H. Duncan 2003. Radar investigations of breaking water waves at low grazing angles with simultaneous high-speed optical imagery. Radio Sci., vol. 38, no. 6 doi:10.1029/2002RS002716.

Stewart, R.H. 1985. Methods of Satellite Oceanography. Berkeley University of California Press, 360 pp.

Stewart, R.H. and J.W. Joy 1974. HF radio measurements of surface currents. Deep Sea Res., vol. 21, no. 12, pp. 1039–1049.

Stofflen, A. and D. Anderson 1997a. Scatterometer data interpretation: Measurement and inversion. J. Atmos. Oceanic Technol., vol. 14, pp. 1298–1313.

Stofflen, A. and D. Anderson 1997b. Estimation and validation of the transfer function CMOD-4. J. Geophys. Res., vol. 102, no. C4, pp. 5767–5780.

Teague, C.C., G.L. Tyler, and R.H. Stewart 1977. Studies of the sea using HF radio scatter. IEEE Trans. Antennas Propag, vol. 26, no. 1, pp. 12–28.

Thompson, D.R. 2004. Microwave scattering from the sea. In Synthetic Aperture Radar Marine User's Manual (ed. by C.R. Jackson and J.R. Apel), Washington, DC: US Department of Commerce, NOAA, September, pp. 117–138.

Thompson, D.R. and B.L. Gotwols 1994. Comparisons of model predictions for radar backscatter amplitude probability density functions with measurements from SAXON. J. Geophys. Res., vol. 99, pp. 9725–9739.

Tikhonov, V.I. 1986. Nonlinear transformations of random processes (in Russian). Radio i sviaz, M., 296 pp.

Tomiyasu, K. 1978. Tutorial review of synthetic aperture radar (SAR) with applications to imaging of the ocean surface. Proc. IEEE, vol. 66, no. 5, pp. 563–583.

Toporkov, J.V., D. Perkovic, G. Farquharson, M.A. Sletten, and S.J. Frasier 2005. Sea surface velocity vector retrieval using dual-beam interferometry: First demonstration. IEEE Trans. Geosci. Remote Sens., vol. 43, no. 11, pp. 2494–2502.

Trizna, D.B. 1985. A model for Doppler peak spectral shift for low grazing angle sea scatter. IEEE J. Oceanic Eng., vol. OE-10, no. 4, pp. 368–375.

Troitskaya, Yu.I. 1994. Modulation of the growth rate of short surface capillary-gravity wind waves by a long wave. J. Fluid. Mech., vol. 273, pp. 169–181.

Troitskaya, Yu.I. 1997. Modulation of the short surface waves riding on a swell wave under the turbulent wind. Quasi-linear model of the growth rate modulation. Ann. Geophys. (Supplement VI to vol. 16. Nonlinear Geophysics & Natural Hazards), C1130.

Tyler, G.L., C.C. Teague, R.H. Stewart, A.M. Peterson, W.H. Munk, and J.W. Joy 1974. Wave directional spectra from synthetic aperture observations of radio scatter. Deep Sea Res., vol. 21, pp. 989–1016.

Ulaby, F.T., R.K. Moore, and A.K. Fung 1986. Microwave Remote Sensing: Active and Passive, vol. 3, Boston MA: Artech House, 1062 pp.

Vachon, P.W. and R.K. Raney 1991. Resolution of the ocean surface wave propagation direction in SAR imagery. IEEE Trans. Geosci. Remote Sens., vol. 29, no. 1, pp. 105–112.

Vachon, P.W. and J.C. West 1992. Spectral estimation techniques for multi-look SAR images of ocean wave. IEEE Trans. Geosci. Remote Sens., vol. 30, no. 3, pp. 568–577.

Valenzuela, G.R. 1978. Theories for the interaction of electromagnetic and oceanic waves – a review. Boundary Layer Meteorol., vol. 13, pp. 61–65.

Valenzuela, G.R. and M.B. Laing 1970. Study of Doppler spectra of radar sea echo. J. Geophys. Res., vol. 75, pp. 551–563.

Vesecky, J.F. and R.H. Stewart 1982. The observation of ocean surface phenomena using imagery from the SEASAT synthetic aperture radar. J. Geophys. Res., vol. 87, no. C5, pp. 3397–3430.

Voronovich, A.G. 1994. Wave Scattering from Rough Surfaces. 2nd ed., New York: Springer Verlag, 228 pp.

Wackerman, C.C. and P. Clemente-Colon 2004. Wave refraction, breaking and other near-shore processes. In Synthetic Aperture Radar Marine User's Manual (ed. by C.R. Jackson and J.R. Apel), Washington, DC: US Department of Commerce, NOAA, September, pp. 177–188.

Walsh, E.J., D.W. Hancock, D.E. Hines, R.N. Swift, and J.E. Scott 1985. Directional wave spectra measured with the surface contour radar. J. Phys. Oceanogr., vol. 15, pp. 566–592.

Walsh, E.J. and D.C. Vandemark 1998. Measuring sea surface mean square slope with a 36-GHz scanning radar altimeter. J. Geophys. Res., vol. 103, no. C6, pp. 12587–12601.

West, J.C., R.K. Moore, and J.C. Holtzman 1990. Synthetic-aperture-radar imaging of the ocean surface using the slightly-rough facet model and a full surface-wave spectrum. Int. J. Remote Sens., vol. 11, no. 8, pp. 1451–1480.

Wetzel, L. 1986. On microwave scattering by breaking waves. In Wave Dynamics and Radio Probing of the Ocean Surface (eds. by Phillips, O.M. and K. Hasselmann), NY: Plenum Press, pp. 273–284.

Wright, C.W., E.J. Walsh, D.C. Vandemark, W.B. Krabill, A.W. Garcia, S.H. Houston, M.D. Powell, P.G. Black, and F.D. Marks 2001. Hurricane directional wave spectrum spatial variation in the open ocean. J. Phys. Oceanogr., vol. 31, pp. 2472–2488.

Wright, J.W. 1968. A new model for sea clutter. IEEE Trans. Antennas Propag., vol. AP-16, no. 2, pp. 217–223.

Zavorotny, V.U. and A.G. Voronovich 1998. Two-scale model and ocean radar Doppler spectra at moderate- and low-grazing angles. IEEE Trans. Antennas Propag., vol. 46, no. 1, pp. 84–92.

Zhydko, Yu.M. and G.K. Ivanova 2001. Wind noise in a radar signal of 3-cm radio waves reflected from the sea surface (in Russian). Izv. VUZ-Radiofizika, vol. 44, no. 8, pp. 653–658.

Zhydko, Yu.M., M.B. Kanevsky, and V.V. Rodin 1983. Wind noise in the *cm* radar signal backscattered from the sea surface (in Russian). Izv. AN SSSR, Fiz. Atm. Okean., vol. 19, no. 3, pp. 328–329.

Zilman, G., A. Zapolski, and M. Marom 2004. The speed and beam of a ship from its wake SAR images. IEEE Trans. Geosci. Remote Sens., vol. 42, no. 10, pp. 2335–2343.

Index

Printed and bound by CPI Group (UK) Ltd, Croydon, CR0 4YY

14/05/2025

01871145-0001